Institute for Nonlinear Science

T0202863

SPRINGER
STUDY
EDITION

Springer

New York
Berlin
Heidelberg
Barcelona
Hong Kong
London
Milan
Paris
Singapore
Tokyo

Institute for Nonlinear Science

Henry D.I. Abarbanel

Analysis of Observed Chaotic Data

With 140 Illustrations

 Springer

Henry D.I. Abarbanel
Professor of Physics
Director, Institute for Nonlinear Science
University of California — San Diego
La Jolla, CA 92093-0402, USA

Library of Congress Cataloging-in-Publication Data
Abarbanel, H.D.I.
  Analysis of observed chaotic data / Henry D.I. Abarbanel.
    p. cm. — (Institute for nonlinear science)
  Includes bibliographical references and index.
  ISBN 0-387-94523-7 (hardcover : alk. paper)
    1. Chaotic behavior in systems. I. Title. II. Series: Institute
  for nonlinear science (Series)
  Q172.5.C45A23 1995
  501′.1385 — dc20   95-18641

Printed on acid-free paper.

Production managed by Frank Ganz; manufacturing supervised by Jacqui Ashri.
Photocomposed pages prepared from the author's LAT_EX file using Springer-Verlag's "svsing.sty" macro.
Printed and bound by Edwards Brothers Inc., Ann Arbor, MI.
Printed in the United States of America.

9 8 7 6 5 4 3 2

ISBN 0-387-94523-7 (Hardcover)
ISBN 0-387-98372-4 (Softcover)   SPIN 10830839

Springer-Verlag is a part of *Springer Science+Business Media*

springeronline.com

Preface

When I encountered the idea of chaotic behavior in deterministic dynamical systems, it gave me both great pause and great relief. The origin of the great relief was work I had done earlier on renormalization group properties of homogeneous, isotropic fluid turbulence. At the time I worked on that, it was customary to ascribe the apparently stochastic nature of turbulent flows to some kind of stochastic driving of the fluid at large scales. It was simply not imagined that with purely deterministic driving the fluid could be turbulent from its own chaotic motion. I recall a colleague remarking that there was something fundamentally unsettling about requiring a fluid to be driven stochastically to have even the semblance of complex motion in the velocity and pressure fields. I certainly agreed with him, but neither of us were able to provide any other reasonable suggestion for the observed, apparently stochastic motions of the turbulent fluid. So it was with relief that chaos in nonlinear systems, namely, complex evolution, indistinguishable from stochastic motions using standard tools such as Fourier analysis, appeared in my bag of physics notions. It enabled me to have a physically reasonable conceptual framework in which to expect deterministic, yet stochastic looking, motions.

The great pause came from not knowing what to make of chaos in nonlinear systems. By this I mean that when I learn about a new phenomenon in Physics, I try to provide myself a framework in which to ask what I have learned about the physical system in which it is observed. In the case of the BCS theory of superconductivity, just to stray a bit, I recall understanding immediately that it represented a deep insight into collective effects in quantum systems leading to remarkable macroscopic properties.

In the case of chaos, I was simply bewildered why nature chose such an apparently unwelcome outcome for the evolution of physical systems. From the outset it was clear, especially for a physicist, from the elegant fluid dynamical experiments of colleagues like Jerry Gollub and Harry Swinney, that chaos was a real property of physical systems, and thus I could not ignore it as an interesting, but irrelevant, mathematical construct. Yet I could not give meaning to why a system would not toe the line and behave as I had learned in my courses on classical physics, namely, in nice regular ways. Perhaps I was ready to accept a superposition of nice regular ways, namely, quasi-periodic behavior composed of a collection of incommensurate sinusoids. I was not ready for evolution which was **nonperiodic** and thus possessed of a continuous, broad Fourier spectrum. So it gave me great pause, and this book in some manner puts forward some of the answers I have put together to move beyond that pause.

While I may still not have a fully acceptable (metaphysical, I guess) answer as to why physical systems choose to behave chaotically, I have some clearer ideas on what is accomplished by their evolving chaotically. I even have some personally satisfying conjectures as to what is achieved in a functional sense for both physical and biological systems by the ability to develop chaotic motion. My understanding rests on the perception that the ability of a system in chaotic motion to explore wide regions of its phase space, and thus utilize information about a large collection of potential states of the system, can have distinct positive implications for what the system can do in achieving its required functions. Regular systems moving on a torus in some dimension explore a really limited domain of phase space. That may be perfectly fine for some actions, but such limited behavior is rather restrictive, and if the systems want to get to another state by action of external forces or equivalently a changed environment, it will be most difficult to do that when the basic motion is confined to an integrable subspace of the full system phase space. Biological systems that do not take advantage of chaos in their state spaces may not be able to participate in evolution to the extent we understand that as adaptation to changing environments. Of course, we are unlikely to see such systems as they are now extinct. Similarly, the range of alternatives available to which one can control a system is larger if chaotic motion, seen as a wide ranging exploration of state space, is allowed.

Perhaps that sounds more like a credo than an understanding, but keeping in mind the possibilities that wide domains of phase space are opened up by chaotic motion focuses one's attention on the main theme of this book: **geometry in phase space provides answers to a large body of questions as to how one can extract useful physical information from observations of chaos**. The thrust of this book is precisely to provide the physicist and other persons encountering chaotic time series, a set of tools for establishing a framework in which clear questions about the source of the chaotic signal can be asked and answered. An alternative title for the book could have been "The Chaotic Mechanic's Tool Kit: An Owner's Manual."

The book starts on a tone unfamiliar in nonlinear dynamics, namely, with an extended example from manufacturing and milling machines. This example is developed with experimental observations from the laboratory of Professor Frank Moon at Cornell. It is meant to place in context the main goal of this work: to provide tools for scientists and engineers who wish to extract useful content from observations of quantities important in their work but for whom time dependence is irregular in appearance. With data from one set of observations we draw conclusions which are unsupported at first but will rest on the material in the remainder of the book. The discussion of this example will be completed much later in the book.

After this I move right into the analysis of signals observed from chaotic motion. Ideas of phase space and its reconstruction from single observed quantities are introduced. Then details are given on time delay phase space reconstruction. Following this I introduce and discuss the notion of invariants of the chaotic motion, namely, those things **not** sensitive to changes in initial condition.

These preliminaries are then used to discuss model making and prediction within the context of chaos. This is not now, and may never be, a subject which we can consider complete in a mathematical sense. This is an area in which it is quite easy to do very well with rather simple tools, and I shall outline how one accomplishes prediction within the limits of the predictability of the source. Indeed it is a most pleasing aspect of the analysis in this whole book that the limits on predictability of the dynamical system can be systematically extracted from observations of the system.

Following model making I address the issue of signal separation in chaotic systems: given a contaminated chaotic signal, how can we separate the contamination from the chaos. Thinking of this as signal separation, rather than noise reduction which some would term it, allows us to go beyond just removing the contamination, for it may be that the contamination itself is precisely what we wish to capture. Next I give some discussion of various methods which have been explored for controlling chaotic motions either back to regular motions or to chaotic motions somehow more desirable than the one handed to us.

Synchronization of chaotic systems is taken up next. This is in many ways related to control. In controlling chaos we seek to move the free running system we are given into more regular or more chaotic motions by adding dynamics to the original system in the form of time varying parameters or external driving of other forms. In synchronization we seek subspaces of the coupled system space in which a special kind of motion which relates the coupled systems takes place.

I will illustrate the ideas as they are introduced with detailed analysis of computer model calculations and with analysis of data from experiments. Typically the model will either be the three degree of freedom Lorenz model from 1963, the two-dimensional Hénon map, the Rössler attractor [Roe76], or the two-dimensional Ikeda map describing a laser beam in a ring cavity. The experimental data will be taken from observations of nonlinear circuits by Tom Carroll and Lou Pecora of the Naval Research Laboratory or experiments by N. F. Rul'kov at the Institute for Nonlinear Science. At the end of the book will be a chapter discussing a variety of other examples of the application of the chaos toolkit which is the bottom line output of the discussion. The topics which will be covered in this chapter will give us the luxury of allowing us to move from analysis of a sequence of ASCII numbers to attempts to use the tools to uncover innovative understanding of the source of the time series. Among the topics we will discuss are (a) chaotic motions of coherent vortex structures in a turbulent boundary layer, (b) chaotic fluctuations of a solid state laser. (In particular I will show how the tools allow us to establish that some of these fluctuations are due to classical chaos and others have their origin in intrinsic quantum fluctuations of the laser processes.), and (c) chaotic motions of the volume of the Great Salt Lake.

In the case of the laser intensity fluctuations I will also discuss a dynamical model which ties together the conclusions drawn from our analysis of a single observed quantity: the intensity of infrared light.

These items represent a broad spectrum of scientific interests and phenomena. They will be discussed in some detail as to their scientific importance, and then I will provide detailed information on what the tools we have developed tell us about these diverse scientific questions. The point of the further examples is to (a) illustrate the power of the methods central to this book, (b) to indicate specific scientific interpretations of aspects of the dynamics of the sources of the various signals, and (c) to place in proper context the practical use of the tools we have so that the reader can see that they do not provide unaided algorithmic paths to correct interpretations of data—humans play a critical role which must be kept in mind.

In the closing chapter we discuss a more abstract topic, namely, the maximum accuracy with which one can expect to ever estimate a dynamical variable of interest when it is chaotic and observed in the presence of external additive noise. This provides a generalization of the usual statement of the implications of the Cramér-Rao bound [Tre68], and ties together some observations in earlier sections, especially surrounding signal separation using manifold decomposition methods.

Despite my original inclinations, I have chosen to forgo the customary introduction to nonlinear dynamics found in most books which have appeared in recent years on topics related to dynamics. At this time there are many excellent introductions and advanced monographs on these topics [Ott93, Dra92, TS86, Moo92, GH83, LL91, Arn78], and the reader is well advised to refer to them when issues of details about strange attractors or other matters arise. I violate this in the Appendix where I discuss two topics which I deem of some importance: (1) the role of information theory in nonlinear systems, and (2) the notion of strange attractors or chaotic behavior arising through a series of instabilities of nonlinear systems. Indeed, the view that motion on a strange attractor is unstable everywhere in state space, yet bounded and possessed of structure in the state space, underlies much of what this book proposes one use in the analysis of such systems. In the case of the Lorenz attractor, seeing some detail about how these instabilities develop and arise from stressing the system to make it perform some task, the transport of heat by convection in the case of the Lorenz model, is worth noting. To make up for the details of these topics I have also included a glossary of terms commonly appearing in this book. This is not a substitute for working through many of the important details to be found in the books I mentioned, but perhaps it will be helpful to the reader just becoming well acquainted with those details to have a handy peek into the meaning of the main items underlying the analysis here. A few sentences describing a strange attractor, for example, can hardly substitute for the reader's working out its properties through examining some examples numerically and reading some of the mathematical results known about such objects, but perhaps the glossary will serve some benefit nonetheless.

It is my hope that the reader will be interested in trying out the methods described in this book on his/her own data and perhaps further developing the analysis tools found here. The algorithms described below are available in a mixture of FORTRAN and C programs. The algorithms have been ported to many computing environments, and a working version running under OSF/Motif on SUN Workstations and other similar UNIX Workstations is available for purchase. Please contact the author. This book is probably the best 'manual' for those programs, though something of a more formal 'here's how to do it' set of directions accompany the software.

The collections of results which make up this book could not have been created by my efforts alone despite the partial hubris required to put them into print above my name alone. I had extensive assistance in putting together the picture painted in this book. Primary among my colleagues have been the graduate students and postdoctoral fellows with whom I have worked over the past several years. Specifically I wish to acknowledge Matt Kennel, Clif Liu, Misha Sushchik, Jesse Goldberg, Garrett Lisi, John J. ("SID") Sidorowich, Reggie Brown, Pierre Marteau, Nikolai Rul'kov, and Paul Bryant. More senior colleagues have also played a critical role, and among them I especially thank Lev Tsimring, Doug Mook, Ed Lorenz, Misha Rabinovich, Cory Myers, and David Ruelle. Ted Frison has supported the effort to put together this book in ways he is too modest to acknowledge. Tom Carroll, Lou Pecora, Upmanu Lall, and Raj Roy have been both friends and invaluable colleagues in providing and interpreting data. Without them this book might seem an amusing set of conjectures; they have made its needed contact with reality.

I have received continuing financial support for the research on which this book could be based. Oscar Manley at the Department of Energy,

Mikhael Ciftan at the Army Research Office, and Michael Shlesinger at the Office of Naval Research have steadfastly given their encouragement to the work here and enabled it to proceed with financial stability. Working with Doug Mook and Cory Meyers of the Lockheed/Sanders Corporation was critical as the algorithms devised here were brought to bear on real world problems.

Over the years I have wondered why authors give such substantial thanks to their families for the support they received while composing scientific volumes. This is no longer a mystery to me as I have so richly benefited from the generosity of my wife Beth Levine and our remarkable daughters Brett and Sara. It is customary to think that one could not have done it without them, but I suspect I have done it so much with them that they could have done it for me. My love and thanks.

October 4, 1995   Henry D. I. Abarbanel
                                                   Del Mar, California

Contents

1
Introduction

We begin our discussions with an example which illustrates the wide range of problems to which the techniques discussed in this book may be applied. The methods we develop as we proceed are quite general and apply to the analysis of time series observations of physical or other systems. The example we have chosen to start the development is comprised of observations of chatter in machine tool production. This is really quite far afield from what one is accustomed to encounter on opening a book on nonlinear dynamics and chaos. I have chosen this example both because it strays from that path and for its intrinsic importance and interest. I ask the reader whose interest in machine tools may be quite minimal to think of this example as illustrative of how one can systematically proceed with the analysis of observed irregular data. It also illustrates important issues about chatter and may have substantial application in terms of making better milling machines. This potential is only one goal of the example.

1.1   Chatter in Machine Tools

In modern manufacturing it is important to make precision cuts in materials and to produce exceptionally smooth surfaces when machining parts of equipment. In consideration of the savings associated with the production of enormous transportation systems such as the Boeing 747 aircraft there have been estimates that if one could assure the production of the majority of the metallic parts of each aircraft to tolerances of $\pm 0.05\,\text{mm}$, nearly

4500 kg of weight would be saved on each vehicle. This would be achieved with no impact at all on performance or safety of operation of the aircraft. Further, smooth surfaces and carefully controlled tolerances would increase the useful lifetime of the equipment and, presumably, the safety to humans during its use.

The most severe limitation in milling operations is often chatter in the tool as it is applied to the piece of material being machined into the desired part. Other factors in determining the limits of performance at fine tolerance can be thermal effects coming from heating while the piece is machined and errors in the precise placement of the tool. These are not of a dynamical origin in the machining process whereby material is selectively sheared from the material, and may be treated as needed by methods outside the discussion here.

Using observations of chatter in a laboratory environment, I will suggest in this chapter that one can utilize those methods to devise innovative control schemes to reduce chatter in these critical manufacturing processes. This discussion will make extensive allusion to the methods developed later in the text. Use of the present methods has not yet moved outside laboratory experiments, though the indications are clear enough to be quite compelling. The success of such an effort would not only lead to cleaner and more precise cutting machines but also would allow the use of those machines at higher speeds further reducing costs for manufacturing. The plan of presentation in this introductory chapter will be to describe the experiments, carried out in the laboratory of Professor F. C. Moon at Cornell University, and proceed as if the reader were slightly familiar with the analysis tools described in the remainder of the book. Of course, we will come back to discuss each item in detail, but the logic of the overall analysis will be stressed from the outset.

1.1.1   The Experiment

Many experiments have been performed by Moon and his students on cutting tools using different tool holders, cutting heads, and workpiece materials. The experiment we discuss was composed of a silicon carbide tool of 0.38 mm nose radius applied on a standard lathe to produce a cut of depth 0.25 mm in aluminum 6061-T6. The tool was fed into the aluminum at the rate of 0.762 mm each revolution. The tool holder was instrumented with strain gauges which read out vertical and horizontal displacements. These were associated with the azimuthal and radial displacements of the tool, respectively, as it impacted the rotating aluminum on the lathe. An illustration of the setup is in Figure 1.1. Data was recorded each $\tau_s = 0.25$ ms, and 20,490 data points, or about 5 s of data, were collected.

In this chapter we examine only the data from the horizontal displacement. In Figure 1.2 we see 4000 of the 20,000 points, namely, data over 1 second of recording. The horizontal displacements are quite irregular with

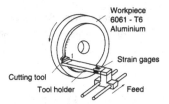

FIGURE 1.1.   Schematic drawing of the cutting tool experiment of F. Moon and students at Cornell University. A workpiece of 6061-T6 aluminum was cut at a depth of 0.25 mm with a silicon carbide tool of 0.38 mm nose radius. The x displacement and the y displacement of the tool holder was registered by strain gauges placed on the holder. The tool was fed into the material at 0.762 mm/revolution. The natural frequency of the tool holder was about 250 Hz, and the data was sampled at 4 KHz ($\tau_s = 0.25$ ms). The machine used was a Hardinge CHNC II CNC Machining Center.

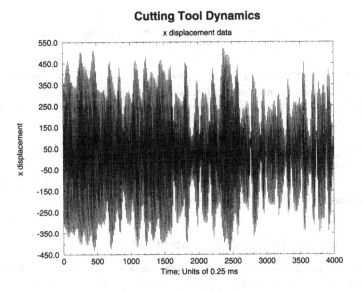

FIGURE 1.2.   Four thousand points (about 1 second) of data from the x displacement of the tool holder. A total of 20,490 points were recorded in this experiment.

a Fourier spectrum as shown in Figure 1.3. In this spectrum there is evidence of a set of peaks, perhaps harmonics of a few fundamental oscillations, with substantial additional power in the broadband background. The broad peak near 250 Hz is associated with the natural frequency of the tool holder. A linear analyst of signals would see narrow spectral lines in these data with contamination by noise. That more than just random high dimensional noise might be part of the process being observed is hinted at, but not yet quantitatively established, by looking at a three-dimensional

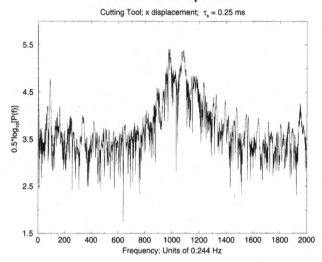

Fourier Power Spectrum

Cutting Tool; x displacement; $\tau_s = 0.25$ ms

FIGURE 1.3.   The Fourier power spectrum associated with the time series in Figure 1.2. The broad peak near 250 Hz is connected with the natural vibrations of the tool holder. The remaining broadband spectrum is the primary interest of this discussion.

portrait of the data as seen in Figure 1.4 where the coordinates of points in the three-dimensional space are $[x(n), x(n+4), x(n+8)]$ at times $n\tau_s$ for $n = 1, 2, \ldots, 1000$. This plot is a reconstructed phase space or state space of the variety we shall discuss extensively in the next few chapters.

1.1.2   Correlations Among Data Points

A key idea we will stress throughout the book is that the space in which to view the structure of the dynamical source of these signals is not that of the one-dimensional observations, here the x(t) x displacements of the tool holder. The dynamics takes place in a space of vectors $\mathbf{y}(t)$ of larger dimension, and we view it projected down on the axis of observed variables. We can identify a space formally equivalent to the original space of variables using coordinates made out of the observed variable and its time delayed copies. This *"phase space reconstruction"* is taken up in the coming chapters. It is to be thought of as a method for providing independent coordinates composed of the present observation $x(t)$, an earlier view of the system $x(t - \tau_1)$ dynamically different from $x(t)$, etc, to produce the ingredients for vectors [ABST93]

$$\mathbf{y}(t) = [x(t), x(t - \tau_1), x(t - \tau_2), \ldots]. \tag{1.1}$$

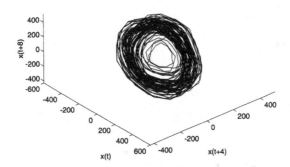

FIGURE 1.4.   A three-dimensional phase portrait of the x displacement of the tool holder using as coordinates for the three dimensional space $[x(t), x(t + 4\tau_s), x(t+8\tau_s)]$. These time delays are determined by examining the average mutual information curve in the next figure.

The choice of values for the τ_i is accomplished by asking when the nonlinear correlation function called average mutual information has its first minimum. This function gives the amount of information, in bits, learned about $x(t+T\tau_s)$, at time $\tau_T = T\tau_s$, from the measurement of $x(t+(T-1)\tau_s)$. The integer multiple T of the sampling time is our goal. The average mutual information among measurements of the x displacement of the cutting holder is seen in Figure 1.5. Here we observe that the first minimum [FS86] is at $T = 4$ or 1 ms. This is how we arrived at the time lags of $T = 4$ and $T = 8$ used in creating the phase portrait in Figure 1.4.

1.1.3   The Number of Coordinates

The number of τ_i to use is determined by asking when the projection of the geometrical structure in Figure 1.4 has been completely unfolded. Unfolding means that points lying close to one another in the space of the $\mathbf{y}(t)$ vectors do so because of the dynamics and not because of the projection. By systematically surveying the data points and their neighbors in dimension $d = 1$, then $d = 2$, etc, we may establish when apparent neighbors stop being "unprojected" by the addition of more components to the $\mathbf{y}(t)$. A plot of the percentage of false neighbors [KBA92] for the x-displacement data of the cutting tool observations is shown in Figure 1.6. From this we can see that at $d = 5$ or $d = 6$, we have unfolded the geometry and find no further unfolding after that. This means that, choosing $d = 6$, or

$$\mathbf{y}(t) = [x(t), x(t + T\tau_s), x(t + 2T\tau_s), \dots x(t + 5T\tau_s)], \qquad (1.2)$$

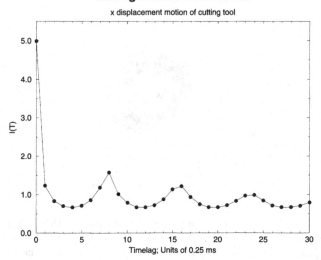

FIGURE 1.5.   The average mutual information for the x displacement of the tool holder. This is a kind of nonlinear correlation function telling how much one learns, in bits, about the measurement $x(t+T\tau_s)$ from the measurement $x(t)$ on the average over all the data. The first minimum of $I(T)$ is at $T = 4$, and using the prescription discussed in Chapter 3 we select this for reconstruction of the state space of the cutting tool dynamics.

we will capture without ambiguity the dynamical features of the source of the x-displacement data. In particular, we will be assured that no unwanted crossings of the orbits defined by the $\mathbf{y}(t)$ will occur and that $\mathbf{y}(t + 1)$ correctly, that is due to the dynamics alone, follows $\mathbf{y}(t)$.

In the present data set this implies that modeling the data in $d = 6$, namely with 6 dynamical degrees of freedom or 6 differential equations governing the evolution of dynamical variables, will be adequate, indeed necessary, for capturing the properties of the source. These variables are related in some unknown nonlinear fashion to dynamical variables such as positions and velocities of parts of the tool holder. As it happens knowledge of that relationship is not required to carry out many tasks of great use including predicting and possibly controlling the motions of the tool holder. If actual equations of motion are available, one may utilize the data to verify their properties in a statistical sense. Comparison of a numerical solution of the equations directly with the data is certain to fail as the data are nonperiodic and unstable everywhere is phase space.

One may additionally inquire whether the number of required degrees of freedom is locally fewer than the global number for unfolding. We will later discuss examples where this is required and how we establish if this is the

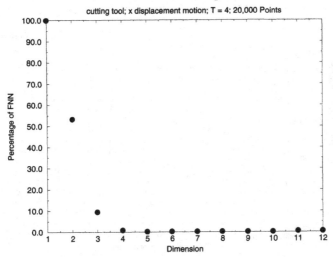

FIGURE 1.6.   The percentage of false nearest neighbors associated with vectors $[x(t+T\tau_s), x(t+2T\tau_s), \ldots, x(t+(d_E-1)T\tau_s)]$ as a function of d_E. The number of false neighbors indicates the degree to which the attractor seen in Figure 1.4 remains unfolded by its projection onto the observation axis $x(t)$. When the number of false neighbors drops to zero, the attractor is unfolded and remains unfolded in higher dimensions. False neighbors are connected with unpredictability as phase space points will be near each other for nondynamical reasons, namely just because of projection. The d_E where the percentage of false nearest neighbors drops to zero is the necessary global dimension to be used to examine the observed data.

case, but for this example it happens that the local number of dynamical variables is also six [AK93].

1.1.4   Prediction of the x displacement of the Cutting Tool

Using our knowledge of the local and global coordinates of the dynamics we can, working in the global space for $\mathbf{y}(t)$, make local models of how each neighborhood, containing only true neighbors now, evolves into other neighborhoods of the orbit $\mathbf{y}(t)$ [ABST93]. The simplest such model establishes a local linear rule

$$\mathbf{y}(t+1) = \mathbf{A}_t + \mathbf{B}_t \cdot \mathbf{y}(t), \tag{1.3}$$

connecting $\mathbf{y}(t)$ to $\mathbf{y}(t+1)$. The coefficients $\mathbf{A}_t$ and $\mathbf{B}_t$ are to be determined by a local least-squares fit to the way the whole neighborhood of $\mathbf{y}(t)$ evolves into the neighborhood of $\mathbf{y}(t+1)$. We can determine such coefficients all along the data set, and then use this information as a lookup table for establishing how a new data point $\mathbf{z}$ will evolve. For this purpose we find

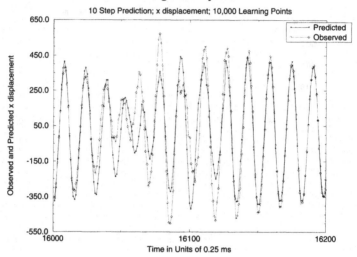

FIGURE 1.7.   Predictions ten steps (of τ_s) ahead on the x-displacement attractor. Local linear models of evolution for the system were constructed using 10,000 learning points from the full 20,490 points of the measured data. Predictions were made starting with point 16,000. These predictions ten steps ahead are compared with the observed data at the same time. The prediction horizon associated with the largest Lyapunov exponent of the system is about $14\tau_s$.

the nearest neighbor $\mathbf{y}(t')$ to $\mathbf{z}$ among the original data set. The point $\mathbf{z}$ evolves to

$$\mathbf{z} \to \mathbf{A}_{t'} + \mathbf{B}_{t'}\mathbf{z}, \tag{1.4}$$

according to this rule. We can then evolve this new point by the same method.

In Figure 1.7 we see the result of using 10,000 of the x-displacement data points to determine coefficients $\mathbf{A}_t$ and $\mathbf{B}_t$, then using the local rules established in this fashion to predict ahead 10 steps, one step at a time, from point 16,000 through point 16,200. The predictions work rather well failing locally at certain regions in the data, then recovering again.

Using the generalization of linear stability theory as applied to chaotic, time dependent orbits such as $\mathbf{y}(t)$, we can determine that errors in precision measurements of the x displacement grow as

$$e^{t/(14\tau_s)}, \tag{1.5}$$

so predicting ahead $10\tau_s$ is actually at the limits of predictability in this system [ABST93, Ose68]. Figure 1.8 exhibits the attempt to predict ahead 50 steps using precisely the same methods and same learning data set.

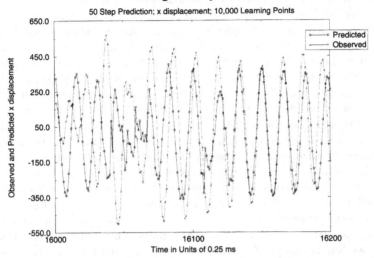

FIGURE 1.8.   Predictions 50 steps (of τ_s) ahead on the x-displacement attractor. Local linear models of evolution for the system were constructed using 10,000 learning points from the full 20,490 points of the measured data. Predictions were made starting with point 16,000. These predictions 50 steps ahead are compared with the observed data at the same time. The prediction horizon associated with the largest Lyapunov exponent of the system is about $14\tau_s$. We are well beyond this horizon here, and the predictions are visibly worse than in Figure 1.7.

While there is occasional success in making these predictions, it is clear that we have gone beyond the predictability horizon by pushing predictions so far. This is good and bad news: the good news is that we have predicted as well as possible according the general theory to be developed in later Chapters. The bad news is that we can reliably predict no further ahead, but that is in the nature of the beast when dealing with chaotic evolution which is unstable everywhere in state space.

1.1.5   Preview of Things to Come

This book is about the methods used to produce the analysis of the x-displacement data just shown. Further, we will discuss how one can use the information just exhibited to produce control strategies for such systems working within the reconstructed phase space itself. This means we will provide the reader with tools to proceed from scalar data $x(t)$ of whatever variable is measured about systems of interest and construct the answers to many interesting and useful questions.

The main focus of this book is to take the mathematical theorems [GH83, Dra92] about the behavior of dynamical systems, when such theorems are available, and combine their insights with those coming from experimental observations of chaotic physical systems and numerical simulations of "toy" systems to develop a systematic methodology for the analysis of experimental data from measurements on physical and biological systems. In the next chapter the subject of reconstruction of phase space from scalar observations is taken up. Following that we discuss choosing time delays, false nearest neighbors and other algorithms for determining the size of reconstructed space, invariants of the chaotic motion useful for classifying dynamical systems, and modeling of observed systems. Noise reduction or signal separation is also discussed as are methods used to control chaos to regular motion. The very interesting issue of synchronization among chaotic systems is taken up before, in a dedicated chapter, several examples are discussed in some detail. A final topic treated in detail is the discussion of how well one can estimate the value of a chaotic sequence of measurements when those observations are contaminated by noise. This is the extension of the Cramér-Rao bound [Tre68] to nonlinear signals, and it gives a rationale to the superb signal separation seen by some methods.

An overview of the material in this book along with a heuristic comparison with the methods used to achieve the same ends in the analysis of linear systems is contained in Table 1.1. This is meant to be suggestive of the tasks one sets forth when analyzing data from an unknown source. Clearly it is important to begin with clean, uncontaminated data. In the study of linear systems contamination is generally defined to be broadband spectral stuff which is filtered out against the desired signal which is narrowband lines characteristic of linear evolution. In nonlinear settings, this won't do as both signal and contamination are broadband, and one will certainly throw the baby out with the bathwater using conventional approaches. The job is that of signal separation, and how well one can do that depends on knowledge of the signals one wishes to separate.

Once clean signals are in hand, and unfortunately this may not be known until some analysis has already been performed, the next important task is establishing a correct multivariate state space or phase space in which to work. In this book we focus on time delay phase space reconstruction as it is simple to achieve and explain as well as being general in its application. It is not always "best" or "optimal". In this space characterizing the source of the observations by invariants of the dynamics is an important goal. In linear systems this means finding the resonant frequencies of the source though various means of spectral analysis. The locations of these frequencies provides the collection of classifiers one needs. If the linear system is driven harder or started at a different time, the power under a given spectral peak may change or the phase of a particular spectral component may shift. Classifying nonlinear systems is less transparent. We concentrate here on fractal dimensions and Lyapunov exponents as they are clearly understandable invariant quantities which have a clean interpretation in terms

Linear signal processing	Nonlinear signal processing
Finding the signal **Signal separation** Separate broadband noise from narrowband signal using spectral characteristics. System known: make matched filter in frequency domain.	*Finding the signal* **Signal separation** Separate broadband signal from broadband "noise" using deterministic nature of signal. System known: use manifold decomposition. Separate two signals using statistics on attractor.
Finding the space **Fourier transforms** Use Fourier space methods to turn differential equations or recursion relations into algebraic forms. $x(n)$ is observed; $x(f) = \Sigma x(n) \exp[i2\pi n f]$ is used.	*Finding the space* **Phase space reconstruction** Time lagged variables form coordinates for a phase space in d_E dimensions: $$\mathbf{y}(n) = [x(n), x(n+T), \ldots,$$ $$x(n + (d_E - 1)T)]$$ d_E and time lag T using mutual information and false nearest neighbors.
Classify the signal Sharp spectral peaks. Resonant frequencies of the system *Quantities independent* *of initial conditions*	*Classify the signal* Invariants of orbits. Lyapunov exponents; various fractal dimensions; linking numbers of unstable periodic orbits *Quantities independent* *of initial conditions*
Make models, predict $x(n+1) = \Sigma c_j x(n-j)$ Find parameters c_j consistent with invariant classifiers—location of spectral peaks.	*Make models, predict* $$\mathbf{y}(n) \to \mathbf{y}(n+1)$$ as time evolution $$\mathbf{y}(n+1) = F\mathbf{y}(n), a_1, a_2, \ldots, a_p]$$ Find parameters a_j consistent with invariant classifiers—Lyapunov exponents, fractal dimensions. Models are in local dynamical dimensions d_L; from local false nearest neighbors. Local or global models.

TABLE 1.1. Comparison of Linear and Nonlinear Signal Processing

of the geometric scaling of the attractor or the predictability of the source. No doubt more important work needs to be done in the arena of classifying dynamical systems, and the work in three dimensions on topological classifiers [MSN*91, FHB*91] is certainly along the right path.

With a state space and classifiers in hand, one may build predictive models for the dynamics. These should be built in the state space identified in earlier work and be consistent with the invariants evaluated in that work. One can achieve quite accurate "black box" prediction schemes on an attractor without ever knowing a stitch about the underlying physics, biology, chemistry, or whatever. While satisfying in a sense, this hardly goes to the usual goal of scientific inquiry which is the uncovering of mechanisms for observed behavior. Models for those mechanisms can now be analyzed in a more complete fashion using the results in this book. The specific orbits resulting from numerical solution of those models cannot be compared with the particular observations one has made since if the orbits were chaotic, they will differ in detail in how they traverse the system attractor due to the instability of the system throughout its phase space. One can require that the invariants, fractal dimension and Lyapunov exponents, and others, associated with the model, correspond to those observed as these are invariant statistical quantities corresponding to aspects of the geometrical object called the attractor. This requirement allows one to determine the parameters which are inevitably in a model and to do it in a manner which may be untraditional in that orbits are not compared, but traditional in many other ways.

Unfortunately there is no algorithmic way to perform the model making and comparison of invariants. Intervention of scientific taste and knowledge is required. Indeed this stage of analysis has rarely been achieved, but we shall present one more or less complete case when we look at the observations of a chaotic solid state laser in one of our later chapters.

The examples we draw on for illustration are from physics and from climate data. The methods are clearly applicable to signals from biological systems, chemical systems, and many others. The choice of examples is not meant to be exhaustive but illustrative.

The reader will find numerous repetitions throughout this book. Formulae are repeated as it appears useful not to distract the reader by making them go back chapters to find a definition. This violates the time honored principle of "punitive pedagogy" which holds that everything is stated once and only once regardless of its importance, but I hope the reader will find the repetition bearable as I trust it will increase the pedagogical value of the text.

2

Reconstruction of Phase Space

2.1 Observations of Regular and Chaotic Motions

Chaos comprises a class of signals intermediate between regular sinusoidal or quasiperiodic motions and unpredictable, truly stochastic behavior. It has long been seen as a form of "noise" because the tools for its analysis were couched in a language tuned to linear processes. The main such tool is Fourier analysis which is precisely designed to extract the composition of sines and cosines found in an observation $s(t)$. Indeed, to a Fourier analyst, chaos is little more than "noise". It is broadband spectrally, but continuous broadband rather than just occupying a broad band of frequency space with a lot of sharp narrowband lines.

To the extent that one can design systems for engineering or other use which perform adequately in a linear regime, then linear analysis is all one needs, and the impetus to develop other tools for signal analysis is absent. Physicists do not have this luxury. Physical systems stressed in interesting ways in the laboratory or in the field don't know they are supposed to remain linear, and usually don't. The need to extract interesting physical information about the dynamics of observed physical systems when they are operating in a chaotic regime would itself be enough to motivate physicists to develop new tools for this analysis. The gap between the development of these tools in the 1980s and 1990s, and the recognition that the need was there by Poincaré and others 80 years before, can probably be attributed to the absence of adequate computational power to analyze the nonlinear dynamics underlying chaotic observations.

Times change, and now we are able to make substantial sense out of observations which result in irregular time behavior. This chapter and the ones to follow are an exposition of tools for that purpose. Many of the words and most of the examples are drawn from physics. This reflects the author's scholarly interests, but the tools are not cognizant of the source of the data, and the methods apply with equal force to data from biological observations, social science data, etc. The state of the art in using these tools is such that one cannot simply acquire data, load it into a canned program, click one's mouse on the appropriate analysis tool, and consider oneself done. There is still much human intervention required, especially in the realistic case that the data is contaminated by other signals. Even with good, clean chaotic data, the interpretation of the outcome of the analyses we will describe here, from average mutual information to local Lyapunov exponents, requires some experience with the meaning of the signals. In other words, analysis of data from nonlinear systems is not algorithmic alone, one must bring some experience and taste to the table. This is good news and bad news, but real news.

The best news is that we do have quite powerful tools, as the dedicated reader will see. These tools allow one to distinguish between irregular chaotic signals which come from a low dimensional source and signals which are best categorized as "noise" because their dimension is very high and stochastic methods would be of great utility in studying them. When the dimension is low, the tools allow us to determine if the signal came from a discrete time map or ordinary differential equations as the dynamical rule. In the same circumstance the tools will allow us to make predictive models and characterize or classify the source of the observations. All this with no knowledge of the mechanisms, physical, biological, or other, that are at work.

When knowledge of those mechanisms is available, then the tools provide a way to check the outcome of those dynamical mechanisms when the output is chaotic. The comparison is on the basis of the statistical evaluations which the tools represent. The tools are primarily geometric and use again and again the notion of neighborhoods in a phase space which itself is reconstructed from the observations. In this sense they are applicable to signals from linear sources, but we have no pretensions that they can compete with the excellent methods developed over the past 50 years or more to deal with such signals.

Much of the development to follow will draw on examples from computer simulations of nonlinear systems or from clean laboratory data from nonlinear electrical circuits operating at low frequencies, so lumped parameter descriptions are excellent. After the various tools have been described we shall use them to study realistic data sets from several sources. We ask the reader's indulgence as we develop the methods before we enter into the interesting, but sometimes uncertain, application to experimental or field data.

2.2 Chaos in Continuous and Discrete Time Dynamics

Chaos occurs as a feature of orbits $\mathbf{x}(t)$ arising from nonlinear evolution rules which are systems of differential equations

$$\frac{d\mathbf{x}(t)}{dt} = \mathbf{F}(\mathbf{x}(t)) \tag{2.1}$$

with three or more degrees of freedom $\mathbf{x}(t) = [x_1(t), x_2(t), x_3(t), \ldots, x_d(t)]$ or invertible discrete time maps [1]

$$\mathbf{x}(t+1) = \mathbf{F}(\mathbf{x}(t)) \tag{2.2}$$

with two or more degrees of freedom [Moo92, GH83, TS86]. Degrees of freedom in systems characterized by ordinary differential equations means the number of required **first order autonomous ordinary differential equations**. In discrete time systems which are described by maps $\mathbf{x}(t) \rightarrow \mathbf{F}(\mathbf{x}(t)) = \mathbf{x}(t+1)$, the number of degrees of freedom is the same as the number of components in the state vector $\mathbf{x}(t)$. The requirement for a minimum size of state space to realize chaos is geometric. For differential equations in the plane (d = 2) it has been known for a long time that only fixed points (time independent solutions) or limit cycles (periodic orbits) are possible. Chaos, as a property of orbits $\mathbf{x}(t)$, manifests itself as complex time traces with continuous, broadband Fourier spectra, nonperiodic motion, and exponential sensitivity to small changes in the orbit.

As a class of observable signals $\mathbf{x}(t)$, chaos lies logically between

i the well studied domain of predictable, regular, or quasi-periodic signals which have been the mainstay of signal processors for decades, and

ii the totally irregular **stochastic** signals we call "noise" and which are completely unpredictable.

> Chaos is irregular in time and slightly predictable.

> Chaos has structure in phase space.

---

[1]Noninvertible maps in one dimension can show chaos as in the example of the logistic map $x \rightarrow rx(1-x)$.

With conventional **linear** tools such as Fourier transforms, chaos looks like "noise", but chaos has structure in an appropriate state or phase space. That structure means there are numerous potential engineering applications of sources of chaotic time series which can take advantage of the structure to predict and control those sources.

2.3   Observed Chaos

From the point of view of extracting quantitative information from observations of chaotic systems, the characteristic features just outlined, pose an interesting challenge to the observer. First of all it is typical to observe only one or at best a few of the dynamical variables which govern the behavior of the system of interest. In the nonlinear circuit examples we often use, usually only a single voltage is measured as the circuit evolves. How are we to go from scalar or univariate observations to the multivariate state or phase space which is required for chaotic motions to occur in the first place?

To address this we focus our attention on discrete time maps. This is really no restriction as in some sense all analysis of physical systems takes place in discrete time: we never sample anything continuously. If we sample a scalar signal $s(t)$ at time intervals τ_s starting at some time t_0, then our data is actually of the form $s(n) = s(t_0 + n\tau_s)$, and the evolution we observe takes us from $s(k)$ to $s(k+1)$.

We can represent continuous flows

$$\frac{d\mathbf{x}(t)}{dt} = \mathbf{F}(\mathbf{x}(t)) \tag{2.3}$$

as finitely sampled evolution

$$\mathbf{x}(t_0 + (n+1)\tau_s) \approx \mathbf{x}(t_0 + n\tau_s) + \tau_s \mathbf{F}(\mathbf{x}(t_0 + n\tau_s)). \tag{2.4}$$

So the observations take

$$\begin{aligned}
s(t_0 + k\tau_s) &\rightarrow s(t_0 + (k+1)\tau_s), \\
s(k) &\rightarrow s(k+1).
\end{aligned} \tag{2.5}$$

There is a special property of flows which we can evaluate with the tools we shall develop in this book. Lyapunov exponents, which we discuss in a later chapter, allow us to determine from finitely sampled scalar data, whether there is a set of ordinary differential equations as the source of that data. With flows a zero global Lyapunov exponent is always present [ABST93, ER85]. For now we lose nothing in generality by concentrating on discrete time evolution of the dynamics.

2.4 Embedding: Phase Space Reconstruction

The answer to the question how to go from scalar observations $s(k) = s(t_0 + k\tau_s)$ to multivariate phase space is contained in the geometric theorem called the embedding theorem attributed to Takens and Mañé [ER85, Man81, Tak81, CSY91]. Suppose we have a dynamical system $\mathbf{x}(n) \rightarrow \mathbf{F}(\mathbf{x}(n)) = \mathbf{x}(n+1)$ where $\mathbf{x}(t)$ phase space is multidimensional. The theorem tells us that if we are able to observe a single scalar quantity $h(\bullet)$, of some vector function of the dynamical variables $\mathbf{g}(\mathbf{x}(n))$, then the geometric structure of the multivariate dynamics can be **unfolded** from this set of scalar measurements $h(\mathbf{g}(x(n)))$ in a space made out of new vectors with components consisting of $h(\bullet)$ applied to powers of $\mathbf{g}(\mathbf{x}(n))$. These vectors

$$\mathbf{y}(n) = [h(\mathbf{x}(n)), h(\mathbf{g}^{T_1}(\mathbf{x}(n))), h(\mathbf{g}^{T_2}(\mathbf{x}(n))), \ldots, h(\mathbf{g}^{T_{d-1}}(\mathbf{x}(n)))], \quad (2.6)$$

define motion in a d-dimensional Euclidian space. With quite general conditions of smoothness on the functions $h(\bullet)$ and $\mathbf{g}(\mathbf{x})$ [CSY91], it is shown that if d is large enough, then many important properties of the unknown multivariate signal $\mathbf{x}(n)$ at the source of the observed chaos are reproduced without ambiguity in the new space of vectors $\mathbf{y}(n)$. In particular it is shown that the sequential order of the points $\mathbf{y}(n) \rightarrow \mathbf{y}(n+1)$, namely, the evolution in time, follows that of the unknown dynamics $\mathbf{x}(n) \rightarrow \mathbf{x}(n+1)$. This is critical in all that follows. The deterministic behavior of the underlying source of observations, $\mathbf{x}(n) \rightarrow \mathbf{x}(n+1)$, assures the deterministic behavior of the substitute representation of this dynamics $\mathbf{y}(n) \rightarrow \mathbf{y}(n+1)$. The integer dimension of the original space need not be the same as the integer dimension of the reconstructed space.

The vector $\mathbf{y}(n)$ is designed to assure that errors in the sequential order which might occur during the projection from the evolution in the original $\mathbf{x}(n)$ space down to the scalar space $h(\mathbf{g}(\mathbf{x}(n)))$ are undone. Such errors result if two points quite far apart in the original space were projected near each other along the axis of scalar observations. This false neighborliness of observations in $h(\mathbf{g}(\mathbf{x}(n)))$ can arise from projection from a higher dimensional space. It has nothing to do with closeness due to dynamics. Further, such an error would be mistaken for some kind of "random" behavior as the deterministic sequence of phase space locations along a true orbit would be interrupted by false near neighbors resulting from the projection.

To implement the general theorem any smooth choice for $h(\bullet)$ and $\mathbf{g}(\mathbf{x})$ is possible [MSN*91]. We specialize here to a choice that is easy to utilize directly from observed data. One uses for the general scalar function $h(\bullet)$ the actual observed scalar variable $s(n)$

$$h(\mathbf{x}(n)) = s(n), \quad (2.7)$$

and for the general function $\mathbf{g}(\mathbf{x})$ we choose the operation which takes some initial vector $\mathbf{x}$ to that vector one time delay τ_s later so the T_k^{th} power of

$\mathbf{g}(\mathbf{x})$ is

$$\mathbf{g}^{T_k}(\mathbf{x}(n)) = \mathbf{x}(n + T_k) = \mathbf{x}(t_0 + (n + T_k)\tau_s), \qquad (2.8)$$

then the components of $\mathbf{y}(n)$ take the form

$$\mathbf{y}(n) = [s(n), s(n + T_1), s(n + T_2), \ldots, s(n + T_{d-1})]. \qquad (2.9)$$

If we make the further useful choice $T_k = kT$, that is, time lags which are integer multiples of a common lag T, then the data vectors $\mathbf{y}(n)$ are

$$\mathbf{y}(n) = [s(n), s(n + T), s(n + 2T), \ldots, s(n + T(d - 1))], \qquad (2.10)$$

composed simply of time lags of the observation at time n τ_s.

These $\mathbf{y}(n)$ replace the scalar data measurements $s(n)$ with data vectors in an Euclidian d-dimensional space in which the invariant aspects of the sequence of points $\mathbf{x}(n)$ are captured with no loss of information about the properties of the original system. The new space is related to the original space of the $\mathbf{x}(n)$ by smooth, differentiable transformations. The smoothness is essential in allowing the demonstration that all the invariants of the motion as seen in the reconstructed time delay space with data $\mathbf{y}(n)$ are the same as if they were evaluated in the original space. This means we can work in the "reconstructed" time delay space and learn essentially as much as we could about the system at the source of our observations as if we were we able to make our calculations directly in the "true" space of the $\mathbf{x}(n)$.

The dimension of the time delay space need not be the same as the original space of the $\mathbf{x}(n)$, and we shall shortly give a rule for a maximum dimension for the time delay space. Constructing this time delay space is quite easy when the data are presented with a common sampling time τ_s. **What time lag $T\tau_s$ to use and what dimension d to use are the central issues of this reconstruction.** The time delay construction using integer multiples of the same lag is the simplest construction and the one which has proven most useful in practice.

The embedding theorem is geometric and addresses the **sufficient** number of components of the vector $\mathbf{y}(n)$ required to assure correct neighborliness of points. It does not prescribe what should be the independent components of the vectors $\mathbf{y}(n)$. There are an infinity of generalizations of the simple construction. If more than one observation is available from an experiment, then use of all observations may prove a good set of coordinates for the phase space vectors $\mathbf{y}(n)$. This means that if we observe data sets $s_1(n)$ and $s_2(n)$, etc, we may use as components of a reconstructed phase space vector $\mathbf{y}(n)$, each of these observations and their time delays until we have enough components. "Enough" will be made quantitative shortly.

2.4.1   Geometry of Phase Space Reconstruction

The basic idea of this construction of a new state space is that if one has an orbit—a time ordered sequence of points in some multivariate space observed at time differences of τ_s—seen projected onto a single axis $h(\bullet)$ or $s(n)$ on which the measurements happen to be made, then the orbit, which we presume came from an autonomous set of equations, may have overlaps with itself in the variables $s(n)$—by virtue of the projection, not from the dynamics. We know there is no overlap of the orbit with itself in the true set of state variables by the uniqueness theorems about the solutions of autonomous equations. Unfortunately we do not know these true state variables, having observed only $s(n)$. If we can "unfold" the orbit by providing independent coordinates for a multidimensional space made out of the observations, then we can undo the overlaps coming from the projection and recover orbits which are not ambiguous.

The simplest example is that of a sine wave $s(t) = A\sin(t)$. Seen in $d = 1$ (the $s(t)$ space) this oscillates between $\pm A$. Two points on this line which are close in the sense of Euclidian or other distance may have quite different values of $\dot{s}(t)$. So two "close" points in $d = 1$ may be moving in opposite directions along the single spatial axis chosen for viewing the dynamics. Seen in two-dimensional space $[s(t), s(t + T\tau_s)]$ the ambiguity of velocity of the points is resolved, and the sine wave is seen to be motion on a figure topologically equivalent to a circle. It is generically an ellipse whose shape depends on the value of T. The overlap of orbit points due to projection onto the one-dimensional axis is undone by the creation of the two-dimensional space. If we proceed further and look at the sine wave in three dimensions, no further unfolding occurs, and we see the sine wave as another ellipse. The unfolded sine wave in two dimensions $(x(n), x(n+2))$ is shown in Figure 2.1 while in three dimensions, it is displayed in Figure 2.2. Figure 2.2 shows the one-dimensional ellipse of the sine wave in coordinates $(x(n), x(n)\cos(\theta) + x(n + 4)\sin(\theta))$, namely as a projection back into two dimensions. This is solely for display. It is clear that once we have unfolded without ambiguity the geometric figure on which the orbit moves, no further unfolding will occur.

The reconstruction theorem recognizes that even in the case where the motion is along a one-dimensional curve, it is possible for the orbit to overlap in points (zero-dimensional objects) when one uses two-dimensional space to view it. If one goes to a three-dimensional space $[s(n), s(n + T), s(n + 2T)]$, then any such remaining points of overlap are undone. The theorem notes that if the motion lies on a set of dimension d_A, which could be fractional, then choosing the integer dimension d of the unfolding space so $d > 2d_A$ is **sufficient** to undo all overlaps and make the orbit unambiguous. As we have seen in the case of the sine wave where $d_A = 1$, it may be possible in particular cases to work with a **necessary** dimension less than the sufficient dimension $d > 2d_A$. We shall describe a way to let the data tell us what necessary value of d to use in order to undo the intersections of an orbit with itself which result from projection of that orbit to lower dimensions.

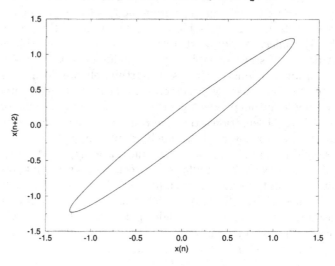

FIGURE 2.1.   The phase space structure of a sine wave seen in two dimensions $(x(n), x(n+2))$ where $x(t) = 1.23\sin(t)$.

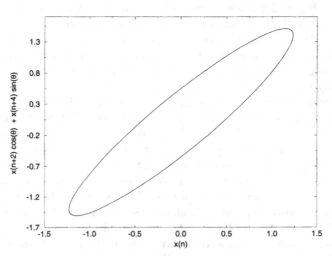

FIGURE 2.2.   The phase space structure of a sine wave seen in three dimensions $(x(n), x(n+2), x(n+4))$ where $x(t) = 1.23\sin(t)$. The figure shows the sine wave in three dimensions projected onto a plane passed through the $x(n)$ axis.

To motivate the essential topological issue in the reconstruction theorem, we note that in a space of dimension d one subspace of dimension d_1 and another subspace of dimension d_2 generically intersect in subspaces of dimension

$$d_1 + d_2 - d. \qquad (2.11)$$

If this is negative, then there is no intersection of the two subspaces. When we ask about the intersection of two subspaces of the same dimension d_A, namely, the orbit with itself, then intersections fail to occur when $2d_A - d < 0$ or

$$d > 2d_A. \qquad (2.12)$$

The question of unfolding a set of dimension d_A from projections to lower dimensions concerns self-intersections of the set with itself as the projections are made. So the relevant criterion for unfolding a single object is $d > 2d_A$. Identifying which among the many possible fractal dimensions should be used as d_A is more difficult [CSY91]. We will address this question after we discuss fractal dimensions.

It is important to note that once one has enough coordinates to unfold any overlaps due to projection, further coordinates are not needed: they serve no purpose in revealing the properties of the dynamics. In principle, they also do no harm unless the orbit is contaminated with some signal of higher dimension than itself. "Noise" will always have this property. "Noise" is formally infinite dimensional and is comprised of sequences of random numbers. It always wishes to be unfolded in a larger dimension than the dynamical signal of interest requires.

This theorem is usually called the embedding theorem as the process of finding a space in which smooth dynamics no longer has overlaps is called embedding. The theorem works **in principle** for any value of T once the dimension is large enough as long as one has an infinite amount of noise free data. This is never going to happen to anyone. This means some thought must be given as to how one may choose both the time delay T and the embedding dimension d when one is presented with real, finite length, and possibly contaminated data. Since a mathematical theorem such as the embedding theorem is silent on the issue of treating real data, the problem is thrown back into the laps of physicists where some intuition and much care is required.

2.5   Reconstruction Demystified

Before our segue into this realism, let us comment on a mystery which may arise in some reader's minds about how measurement of a single scalar could possibly contain enough information to reconstruct a d-dimensional vector space of orbits. The key is that all variables are generically connected in a nonlinear process. Typically there is no disconnected subspace of variables nor can one be found by a smooth transformation of the coordinates. The measurements $s(n)$ and $s(n + T)$ are related by the evolution of the dynamical system over a period T during which all dynamical variables affect the observed variable $s(n)$. So $s(n + T)$ is generically some complicated, unknown, nonlinear combination of all variables of the system. It alone may stand as a proxy for any one of those variables. The combination into a d-dimensional vector of the time delays of $s(n)$ stands as a proxy for d of the active variables of the system being observed.

Another version of this view of embedding or attractor unfolding is to recall from discussions of phase plots in the study of differential equations [Arn82], that if one were able to plot a large number of time derivatives of the observed variable $s(t) : [s(t), \dot{s}(t), \ddot{s}(t), \ldots, \frac{d^N s(t)}{dt^N}, \ldots]$ against one another with points parametrically labeled by t, then we would eventually have provided enough coordinates to recapture the full state space. We cannot know the time derivatives accurately unless the sampling time τ_s is very small, but the approximations

$$\frac{ds(t)}{dt} \approx \frac{s(t + \tau_s) - s(t)}{\tau_s},$$

$$\frac{d^2 s(t)}{dt^2} \approx \frac{s(t + 2\tau_s) - 2s(t + \tau_s) + s(t)}{2\tau_s^2}, \qquad (2.13)$$

and similar approximations to the higher derivatives of $s(t)$, reveal clearly that the new information in each approximate derivative is just the set of time delays we have suggested using for the coordinates of phase space. This idea was apparently first clearly understood by Ruelle [ER85] and the theorems noted were proved subsequently in some generality. In a nonlinear system the coupling of all variables obviates the necessity of making accurate estimates of the time derivatives to capture a set of coordinates for viewing the evolution of the system. This would also be true of any linear system where the linear evolution operator could not be decomposed into actions which allowed independent evolution in subspaces.

Actually there is a trap in trying to make accurate evaluations of the time derivatives of $s(t)$ since each derivative requires one to evaluate various differences between measured quantities [PCFS80]. Taking differences between imperfectly known observations leads quickly to replacement variables which emphasize the noise or error in the original observations. To properly capture small differences between large numbers, one must know those large numbers very accurately. Fortunately one does not need the derivatives of $s(t)$ as coordinates in order to proceed in quite a satisfactory fashion. Recognizing the attempt to evaluate the derivatives themselves as a form of high pass linear filter applied to the data suggests that in doing this operation we emphasize those parts of the data we know the least about, namely, the aspects of the data most contaminated by "noise" or other high frequency stuff.

The main message of this discussion is that starting with a set of scalar observations $s(t_0 + n\tau_s) = s(n)$ we can, using $s(n)$ and its time delays $s(N + T)$, build up vectors in d-dimensional space

$$\mathbf{y}(n) = [s(n), s(n+T), s(n+2T), \ldots, s(n+(d-1)T)] \qquad (2.14)$$

which have no false crossings of the orbits $\mathbf{y}(n)$ because of projection from a higher dimensional space. We have discussed a **sufficient** condition $d > 2d_A$ for the number of entries in $\mathbf{y}(n)$, and we now turn both to a rationale for choosing the time lag T and a **necessary** number of dimensions d.

3
Choosing Time Delays

3.1 **Prescriptions** for a Time Delay

The embedding theorem [Man81, Tak81, CSY91] is silent on the choice of time delay to use in constructing d-dimensional data vectors. Indeed, it allows **any** time delay, except certain multiples of the precise period of a periodic signal [CSY91], as long as one has an infinite amount of infinitely accurate data. Since none of us will ever acquire an infinite amount of infinitely accurate data and still have time to compute anything else, this means one must turn to some principle other than the same geometric construction used in the unfolding of the attractor to establish a workable value for the time delay T. So that absent another theorem—and such a theorem is so far absent—we must find some **prescription** for choosing T. To put it another way, we have no hope of finding the very best or optimal time delay without giving some additional rationale for that optimality. Any such rationale will, of course, determine some T, but such optimality is tautological. At best it simply confirms the prescription one establishes for its realization.

It seems to me best to look at what one wishes to achieve by the choice of time delay.

   i It must be some multiple of the sampling time τ_s, since we only have data at those times. An interpolation scheme to get "more" data is just as uncertain as estimating the time derivatives of $s(t)$.

   ii If the time delay is too short, the coordinates $s(n)$ and $s(n + T)$ which we wish to use in our reconstructed data vector $\mathbf{y}(n)$ will not be independent enough. That is to say that not enough time will have evolved for the system to have explored enough of its state space to produce, in a practical numerical sense, new information about that state space. As a slightly far fetched example which perhaps makes the point, consider sampling the atmosphere every nanosecond. Since the atmosphere does not evolve in any interesting way on such a time scale, we certainly will have substantially oversampled the data and learned essentially nothing for our efforts. If we were to choose $T = \tau_s = 1\,\text{ns}$, we will not have seen any of the dynamics unfold in that time, and we should expect from a physical point of view that $s(n)$ and $s(n + T)$ are for all practical purposes the same measurement.

iii Finally, since chaotic systems are intrinsically unstable, if T is too large, any connection between the measurements $s(n)$ and $s(n+T)$ is numerically tantamount to being random with respect to each other. Even very accurate determinations of the value of $s(n)$ cannot prevent the exponential growth of small errors characteristic of chaos from decorrelating it from the measurement T steps later, when T becomes large.

So what we want is some prescription which identifies a time delay which is large enough that $s(n)$ and $s(n + T)$ are rather independent, but not so large that they are completely independent in a statistical sense. This is not a precise demand on the data. Our approach is to base our choice of T on a fundamental aspect of chaos itself, namely, the generation of information. Thus we will have confidence that our choice is founded in some property of the system we wish to describe. Stable linear systems generate zero information, so selecting information generation as the critical aspect of nonlinear systems assures us from the outset that the quantity we focus on is a property of nonlinear dynamics not shared by linear evolution.

3.2   Chaos as an Information Source

In a system undergoing chaotic motion two nearby points in the state space move exponentially rapidly apart in time. This exponentially rapid separation of orbits in time, often called "sensitivity to initial conditions", is the precise manifestation of the instability everywhere in phase space which leads to the nonperiodic motions we call chaos.

Suppose we have only a fixed, finite resolution in our phase space. For example, imagine we cannot resolve points located within a ball of size R. The limits on resolution may be instrumental or environmental or may just be due to the actual A/D converter with which we acquire the data for storage and analysis. This resolution ball is fixed in time by the setup of the observations. Two points $\mathbf{x}^1(t)$ and $\mathbf{x}^2(t)$ located within this ball at time t cannot be distinguished by our measurements. At some time t' later, the distance between the points has typically grown to $|\mathbf{x}^1(t') - \mathbf{x}^2(t')| \approx |\mathbf{x}^1(t) - \mathbf{x}^2(t)| \exp[\lambda|t'-t|]; \lambda > 0$. When this distance exceeds R, we can then experimentally distinguish the orbits $\mathbf{x}^1(t')$ and $\mathbf{x}^2(t')$, so the instability has uncovered information about the population of state space which was not available at earlier times. There were two orbits within a ball of radius R some time ago, and now we can observe that fact. The rate of information generation can be made precise in terms of an "entropy" due to Kolmogorov and Sinai [Kol58, Sin59]. This can be related via ergodic theorems for these nonlinear systems to the rate of growth of distances by the work of Pesin [Pes77]. An early reference where this idea is explained in physical terms is by Rabinovich [Rab78] who also describes the notion of production of information in a wide variety of physical systems.

This heuristic discussion suggests that the generation of information may be of some fundamental interest in the study of nonlinear dynamical systems. If our system were globally linear and stable, all exponents such as λ are zero or negative, and no information production occurs. If any of the $\lambda > 0$, the linear system is unstable, and the very description of the system as globally linear breaks down. In a nonlinear system of d degrees of freedom, there are d such exponents $\lambda_1 > \lambda_2, \ldots, > \lambda_d$. **Their sum is negative** as that dictates how volumes contract in the realistic dissipative systems we consider. This means that even if only the largest λ_1, is positive, points on the orbit do not run off to infinity despite the implied instability. The nonlinear properties of the dynamics folds the orbit back into a compact region of state space whose volume shrinks at a rate dictated by the sum of all exponents. Positive values of λ stretch the orbits through the implied instability, but the dissipation in the system which bounds the amplitude of motions to be consistent with a finite energy in the system brings the stretched or unstable orbits back into a compact region of state space. This stretching due to instabilities and folding due to dissipation is the fundamental conjunction of processes which gives rise to fractal sets on which the motion takes place and underlies the nonperiodic nature of the chaotic development of a dynamical system [Sma67].

We will argue that for differential equations, systems which evolve in continuous time, one of these instability exponents must be zero. With this we can see how chaos in solutions of differential equations requires three dimensions to occur: one exponent is positive so that chaos is possible, one is zero because we have a differential equation as the source of motions, and a third exponent must exist so that the sum of all three is negative. Three dimensions is the minimum for this, but of course one may have a larger space. For iterated maps no zero exponent occurs in general so one positive exponent is all we require and then one negative exponent so that the sum of exponents is negative and phase space volumes will contract.

3.3   Average Mutual Information

Fundamental to the notion of information among measurements is Shannon's idea of mutual information [Gal68] between two measurements a_i and b_j drawn from sets A and B of possible measurements. The **mutual information** between measurement a_i drawn from a set $A = \{a_i\}$ and measurement b_j drawn from a set $B = \{b_j\}$ is the amount learned by the measurement of a_i about the measurement of b_j. In *bits* it is

$$\log_2\left[\frac{P_{AB}(a_i, b_j)}{P_A(a_i)P_B(b_j)}\right],\tag{3.1}$$

where $P_{AB}(a, b)$ is the joint probability density for measurements A and B resulting in values a and b. $P_A(a)$ and $P_B(b)$ are the individual probability densities for the measurements of A and of B. In a deterministic system we evaluate these "probabilities" by constructing a histogram of the variations of the a_i or b_j seen in their measurements.

If the measurement of a value from A resulting in a_i is **completely independent** of the measurement of a value from B resulting in b_j, then $P_{AB}(a, b)$ factorizes: $P_{AB}(a, b) = P_A(a)P_B(b)$ and the amount of information between the measurements, the mutual information, is zero, as it should be. The average over all measurements of this information statistic, called the **average mutual information** between A measurements and B measurements, is

$$I_{AB} = \sum_{a_i, b_j} P_{AB}(a_i, b_j) \log_2\left[\frac{P_{AB}(a_i, b_j)}{P_A(a_i)P_B(b_j)}\right].\tag{3.2}$$

This quantity is not connected to the linear or nonlinear evolution rules of the quantities measured. It is strictly a set theoretic idea which connects two sets of measurements with each other and establishes a criterion for their mutual dependence based on the notion of information connection between them. We will use this connection to give a precise definition to the notion that measurements $s(t)$ at time t are connected in an information theoretic fashion to measurements $s(t + T)$ at time $t + T$.

So we take as the set of measurements A the values of the observable $s(n)$, and for the B measurements, the values of $s(n+T)$. Then the average mutual information between these two measurements, that is, the amount (in bits) learned by measurements of $s(n)$ through measurements of $s(n+T)$ is

$$I(T) = \sum_{s(n),s(n+T)} P(s(n), s(n + T)) \log_2\left[\frac{P(s(n), s(n + T))}{P(s(n))P(s(n + T))}\right].\tag{3.3}$$

By general arguments [Gal68] $I(T) \geq 0$. $I(T = 0)$ is directly related to the Kolmogorov-Sinai entropy [FS86]. When T becomes large, the chaotic behavior of the signal makes the measurements $s(n)$ and $s(n + T)$ become independent in a practical sense, and $I(T)$ will tend to zero.

It was the suggestion of Fraser [FS86, fra89] that one use the function $I(T)$ as a kind of nonlinear autocorrelation function to determine when the values of $s(n)$ and $s(n+T)$ are independent enough of each other to be useful as coordinates in a time delay vector but not so independent as to have no connection with each other at all. The actual **prescription** suggested is to take the T where the first minimum of the average mutual information $I(T)$ occurs as that value to use in time delay reconstruction of phase space. The detailed suggestion requires the generalization of the notion of average mutual information to higher dimensional phase space where $s(n)$ is replaced by d-dimensional vectors, and one inquires into the marginal increment in information due to the addition of another component of that vector. For essentially all practical purposes one is seeking a value of T which works and is connected with the nonlinear information generating properties of the dynamics. The simple rule for $I(T)$ as just defined serves quite well.

The choice of the first minimum of the average mutual information is reminiscent of the choice of the first zero of the linear autocorrelation function

$$C(T) = \sum_n [s(n) - \bar{s}] \, [s(n + T) - \bar{s}],$$

$$\bar{s} = \frac{1}{N} \sum_{n=1}^{N} s(n), \tag{3.4}$$

to be the time at which one chooses the lag T. This choice is the optimum **linear** choice from the point of view of predictability, in a least squares sense, of $s(n+T)$ from knowledge of $s(n)$. What such a linear choice has to do with the nonlinear process relating $s(n)$ and $s(N+T)$ is not at all clear, and we cannot recommend this choice at all. Indeed, we will give an example in one of the later chapters where the first minimum of the autocorrelation function is completely misleading as a guide for the choice of the time lag T. We illustrate the idea of average mutual information by displaying its values from a standard chaotic system, the Lorenz model [Lor63], and then from the physical system which we will discuss as an example throughout this book.

3.3.1   Lorenz Model

The system of three ordinary differential equations abstracted by Lorenz in 1963 from the Galerkin approximation to the partial differential equations of thermal convection in the lower atmosphere derived by Salzman [Sal62] now stand as a workhorse set of equations for testing ideas in nonlinear

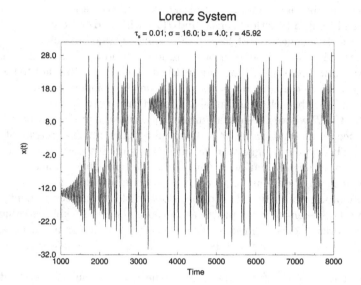

FIGURE 3.1.   The time series of the $x(t)$ component from the Lorenz model, Equation (3.5). The parameters used in the model were $\sigma = 16.0, b = 4.0$ and $r = 45.92$. In the Runge-Kutta integration routine a time step $\tau_s = 0.01$ was used.

dynamics. The equations read

$$
\begin{aligned}
\dot{x} &= \sigma(y - x). \\
\dot{y} &= -xz + rx - y, \\
\dot{z} &= xy - bz,
\end{aligned}
\tag{3.5}
$$

and we use the standard parameter values $\sigma = 16$, $b = 4$ and $r = 45.92$. The three state variables are scaled and dimensionless amplitudes of two components of temperature and one component of velocity in the underlying convection problem. Convection occurs when $r \approx 1$, so we are using the equations in a domain far from a connection with physics. Nonetheless, they serve as a very useful arena for studying dynamics. In this parameter regime the orbits of the Lorenz system reside on a geometric object of dimension $d_A \approx 2.06$ and exhibit nonperiodic, chaotic motion.

These equations were solved using a straightforward fourth order Runge-Kutta method with time step of $\tau_s = 0.01$. This is smaller than necessary to capture the essence of the dynamics of this system, but is chosen to assure stability and accuracy in the integration of the equations with this simple, easily programmed algorithm. In these dimensionless time units one visits the entire attractor once in about 0.5 time units, so 50 time steps gets us around the attractor about once. The time series for these equations is very familiar; it is displayed in Figure 3.1. We display in Figure 3.2 the

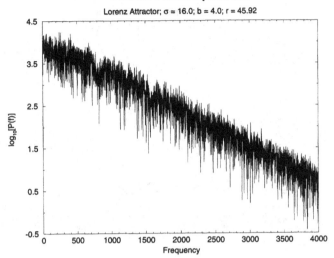

Fourier Power Spectrum

Lorenz Attractor; σ = 16.0; b = 4.0; r = 45.92

FIGURE 3.2.   The Fourier power spectrum of the signal shown in Figure 3.1. The spectrum is continuous and broadband which is what we expect of a nonperiodic signal.

Fourier power spectrum of the signal $x(t)$ from the Lorenz system, and in Figure 3.3 the autocorrelation function. The spectrum is quite unrevealing as we might expect from a nonperiodic signal. The first zero of the autocorrelation function is at $T \approx 30$.

Next we evaluate the average mutual information $I(T)$ from the time series for $x(n) = x(t_0 + n\tau_s)$. To determine $P(x(n))$ we simply project the values taken by $x(n)$ versus n back onto the $x(n)$ axis and form a normalized histogram of the values. This gives us $P(x(n))$. For the joint distribution of $x(n)$ and $x(n + T)$ we form the two-dimensional normalized histogram in the same way. These operations are done easily and quickly on a standard workstation. The result is the curve for $I(T)$ displayed in Figure 3.4. We see quite clearly that the first minimum occurs at $T = 10$.

Now what happens if we were to choose $T = 9$ or even $T = 8$? Nothing really. Since the selection of the first minimum of $I(T)$ is a prescription, this means that for values of T near this minimum the coordinate system produced by time delay vectors is essentially as good as that for the T which is the actual first minimum of $I(T)$. The visual picture of the reconstructed attractor for systems such as the Lorenz attractor or other low dimensional systems changes gracefully for variations around the minimum of $I(T)$ [FS86]. Choosing the actual first minimum appears a useful and workable prescription.

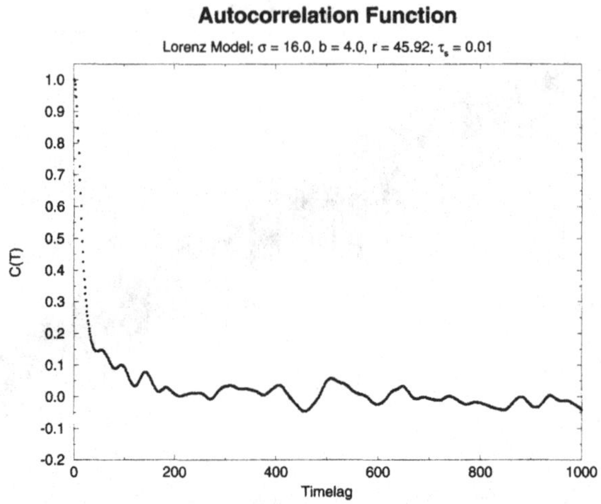

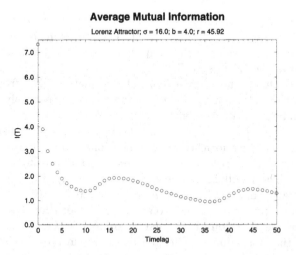

Autocorrelation Function

Lorenz Model; σ = 16.0, b = 4.0, r = 45.92; τ$_s$ = 0.01

FIGURE 3.3.   The standard linear autocorrelation function for the signal $x(t)$ in Figure 3.1. If we used the first zero crossing of this autocorrelation function to select the time delay T for time delay embedding, we would choose $T \approx 30$.

Average Mutual Information

Lorenz Attractor; σ = 16.0; b = 4.0; r = 45.92

FIGURE 3.4.   The average mutual information $I(T)$ for the signal $x(t)$ in Figure 3.1. The first minimum of this function is at $T = 10$, and this is the value of time lag we utilize for phase space reconstruction. It corresponds to approximately 50 trips around the attractor.

Choosing one T or another is tantamount to choosing different coordinate systems in which the data vectors $\mathbf{y}(n)$ are evaluated. The connections among these different coordinate systems, while unknown to us, are assumed to be smooth and differentiable. When we evaluate properties of the data seen in one coordinate system or another, we should be careful that the values determined are not sensitive to the choice of coordinates, namely, to the choice of T.

3.3.2 Nonlinear Circuit with Hysteresis

In order to illustrate the various analysis tools we discuss throughout this book we have selected data from a nonlinear circuit built by Lou Pecora and Tom Carroll at the U. S. Naval Research Laboratory [CP91]. The circuit has three loops with a nonlinear element exhibiting hysteresis and was designed to be used in demonstrating synchronization of chaotic systems. We will come to synchronization later, but for now we have a nice, clean source of chaotic data from a system which we have great confidence in characterizing by three equations arising from Kirchoff's circuit laws. It is perhaps a mark of the times that I have never seen this circuit, and the data arrived in my office through the Internet. In that sense it is an excellent example for our discussions here as this is equivalent to a single observation of some system without the observer having the ability to tinker with the source of the data to produce some desired effect.

A detailed description of the components and properties of the hysteretic circuit used for our nonlinear circuit data can be found in [CP91]. The circuit consists in essence of an unstable second degree oscillator, which operates in the range of several hundred Hertz, connected to an hysteretic element. The circuit is similar to some described by Newcomb and his collaborators [NS83].

The circuit is modeled by the three ordinary differential equations

$$\frac{dx_1(t)}{dt} = x_2(t) + \gamma x_1(t) + c x_3(t),$$

$$\frac{dx_2(t)}{dt} = \omega x_1(t) - \delta_2 x_2(t),$$

$$\epsilon \frac{dx_3(t)}{dt} = (1 - x_3(t)^2)(S x_1(t) - D + x_3(t)) - \delta_3 x_3(t) \qquad (3.6)$$

with $\gamma = 0.2, c = 2.2, \delta_2 = 0.001, \delta_3 = 0.001, \epsilon = 0.3, \omega = 10.0, S = 1667.0$, and $D = 0.0$. The constants in the equation for $x_3(t)$ were set to give a square hysteretic loop.

We analyzed two voltages measured in this circuit. The first which we call $V_A(t)$ was taken from $x_1(t)$ while the second $V_B(t)$ was taken from $x_3(t)$. $\tau_s = 0.1$ ms for these measurements, and 64,000 data points were taken for each voltage. We will discuss the analysis of both sets of data in

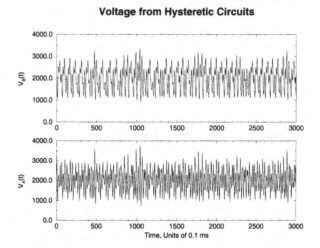

FIGURE 3.5.   The time series for voltages $V_A(t)$ and $V_B(t)$ from the nonlinear circuit with an hysteretic element. The data is from L. M. Pecora and T. L. Carroll at the U. S. Naval Research Laboratory. 64,000 data points from each tap into the circuit were provided. The sampling time is $\tau_s = 0.1\,\text{ms}$.

what follows. This is done to illuminate the many remarks made explicitly and implicitly in this monograph about one's ability to learn about a dynamical system from a single measurement. We should be able to learn the same things, then, from either measurement. Also we will see that the two measurements require slightly different specifications in terms of time delay phase space reconstruction in places where these differences are allowed by the general ideas, but where the results must agree, they do.

We show in Figure 3.5 the time traces of the two voltages and in Figure 3.6 the Fourier power spectrum for V_A, and in Figure 3.7, the spectrum for V_B. As we expect these figures remind us that the time trace is complex and nonperiodic and the same reminder is present in the Fourier spectrum. In Figure 3.8 we display the $I(T)$ for both voltages. Each has its first minimum at $T = 6$, and we shall use that value in subsequent calculations as we proceed to further analyze the circuit using the tools we shall build up as we go along.

3.4   A Few Remarks About $I(T)$

Average mutual information $I(T)$ reveals quite a different insight about the nonlinear characteristics of an observed time series than the more familiar linear autocorrelation function. The latter is tied to linear properties of

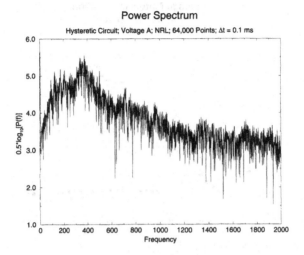

FIGURE 3.6. The Fourier power spectrum for the voltage $V_A(t)$ from the nonlinear circuit time series shown in Figure 3.5. The Fourier spectrum has a broad peak and a continuous, broadband spectrum indicative of nonperiodic behavior.

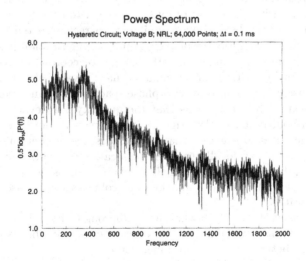

FIGURE 3.7. The Fourier power spectrum for the voltage $V_B(t)$ from the nonlinear circuit time series shown in Figure 3.5. The Fourier spectrum has a broad peak and a continuous, broadband spectrum indicative of nonperiodic behavior.

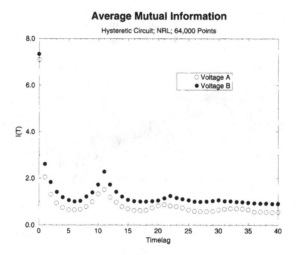

Average Mutual Information

Hysteretic Circuit; NRL; 64,000 Points

FIGURE 3.8.   The average mutual information for the two voltages $V_A(t)$ and $V_B(t)$ from the nonlinear circuit with hysteretic element. Each voltage has its first minimum of $I(T)$ at $T = 6$ or at a time equal to 0.6 ms.

the source. Fortunately the average mutual information between measurements $s(n)$ and time lagged measurements $s(n+T)$ is both easy to evaluate directly from the time series $s(n)$ and easy to interpret. There should be no barrier to using it instead of autocorrelation to establish a reasonable choice for the time lag to use in building up reconstructed phase space vectors. A very nice property of average mutual information [Gal68, FS86] is that it is invariant under smooth changes of coordinate system. This is easily seen by writing the formula we have given which is couched in terms of discrete time in terms of a phase space integral and then changing variables explicitly. This means that the quantity $I(T)$ evaluated in time delay coordinates and in the original, but unknown, coordinates takes on equal values. It also suggests that in the presence of contaminated measurements, which act locally to alter the location of points and the details of the local distributions entering the probability densities involved in the evaluation of $I(T)$, that $I(T)$ may be more robust against noise than many other quantities.

If one wishes to be somewhat more thorough about the time lag to choose [fra89], it is possible to ask a broader question: what is the mutual information between a vector in dimension d

$$\mathbf{y}(n) = [s(n), s(n+T), \ldots, s(n+(d-1)T)], \tag{3.7}$$

and an additional component $s(n+Td)$? This would be computed using the formula for mutual information we gave earlier, Equation (3.2), now choosing the set of measurements A to be the collection of d-dimensional vectors $\mathbf{y}(n)$ and the measurements B the measurements $s(n + dT)$. This would gives us the answer to the question, how much is learned about the measurement $s(n+dT)$ from the measurements of $\mathbf{y}(n)$ on the average over the attractor? If this number were small, relative to the value at $T = 0$ for example, then we would conclude the d-dimensional vector $\mathbf{y}(n)$ was capturing the independent coordinates of the source using the time lag T. The additional component $s(n + dT)$ would add little or no additional information to that contained in $\mathbf{y}(n)$. A whole series of mutual information like functions are available in higher dimensions, but have seen little use in analysis of real data as the requirements on the amount of data for their accurate estimate grows exponentially with dimension.

If we may once again emphasize that the choice of T is by prescription, one must examine the practical utility of evaluating these higher order information theoretic quantities given their computational requirements. if there is some new aspect of the nonlinear dynamics revealed by this, it may be worth the effort.

There is numerical evidence from a paper of Vastano and Swinney [VS89] that the mutual information statistic is much more sensitive to detection of chaotic motions in a background of other activity than traditional correlation functions composed of powers of the dynamical variables. The use of mutual information produced a clear signal of propagating information in a lattice of coupled nonlinear oscillators while polynomial correlation functions of orders up to quartic showed essentially no signal. While we do not pursue this facet of mutual information, it is one more indication that information is a key property of nonlinear systems which we should exploit to learn about their inner workings.

4

Choosing the Dimension of Reconstructed Phase Space

4.1 Global Embedding Dimension d_E

We wish to determine the integer global dimension where we have the necessary number of coordinates to unfold observed orbits from self overlaps arising from projection of the attractor to a lower dimensional space. For this we go into the data set and ask when the unwanted overlaps occur. The lowest dimension which unfolds the attractor so that none of these overlaps remains is called the embedding dimension d_E. d_E is an integer. If we measure two quantities $s_A(n)$ and $s_B(n)$ from the *same* system, there is no guarantee that the global dimension d_E for from each of these is the same. Each measurement along with its timelags provides a different nonlinear combination of the original dynamical variables and can provide a different global nonlinear mapping of the true space $\mathbf{x}(n)$ into a reconstructed space of dimension d_E where smoothness and uniqueness of the trajectories is preserved. Recall that d_E is a **global** dimension and may well be different from the local dimension of the underlying dynamics.

To illustrate this important point imagine a two-dimensional mapping which globally evolves on a Möbius strip. The global dimension to unfold the dynamics would be $d_E = 3$ while the local dimension would be two, of course. We shall come to local or dynamical dimensions; the local dimension revealed by two different measured variables must be the same. For now we focus on global embedding or unfolding dimension alone.

The embedding theorem tells us that if the dimension of the attractor defined by the orbits is d_A, then we will certainly unfold the attractor in an integer dimensional space of dimension d_E where $d_E > 2d_A$. This is not the necessary dimension for unfolding, but is sufficient and certainly tells us when to stop adding components to the time delay vector.

The fractal dimension, in this case the box counting dimension [CSY91], of the strange attractor for the Lorenz model is $d_A \approx 2.06$ which would lead us to anticipate $d_E = 5$ to unfold the Lorenz attractor. As it happens $d_E = 3$ will do for this system when we use as the observations $x(t)$. We now give the arguments and later show the calculations which lead to this result. Let us emphasize again that $d_E = 3$ is not to be associated with the original three-dimensional nature of the differential equation since we are utilizing coordinates $[x(n), x(n+T), x(n+2T)]$ which are nonlinearly related to the original $[x(t), y(t), z(t)]$. We are certain from the theorem that some dimension ≤ 5 will do for d_E, but we will now discuss which dimension to select.

4.2   Global False Nearest Neighbors

Suppose we have made a state space reconstruction in dimension d with data vectors

$$\mathbf{y}(k) = [s(k), s(k+T), \ldots, s(k+(d-1)T)], \tag{4.1}$$

using the time delay suggested by average mutual information. Examine the nearest neighbor in phase space of the vector $\mathbf{y}(k)$ with time label k. This will be a vector

$$\mathbf{y}^{NN}(k) = [s^{NN}(k), s^{NN}(k+T), \ldots, s^{NN}(k+(d-1)T)], \tag{4.2}$$

and its time label bears little relation to the time k at which the vector $\mathbf{y}(k)$ appears. If the vector $\mathbf{y}^{NN}(k)$ is truly a neighbor of $\mathbf{y}(k)$, then it came to the neighborhood of $\mathbf{y}(k)$ through dynamical origins. It is either the vector just ahead or just behind $\mathbf{y}(k)$ along the orbit, if the time steps along the orbit are small enough, or it arrived in the neighborhood of $\mathbf{y}(k)$ through evolution along the orbit and around the attractor. Since the attractor for real physical systems is quite compact in phase space, each phase space point will have numerous neighbors as the number of data becomes large enough to populate state space well.

If the vector $\mathbf{y}^{NN}(k)$ is a **false** neighbor of $\mathbf{y}(k)$ having arrived in its neighborhood by projection from a higher dimension because the present dimension d does not unfold the attractor, then by going to the next dimension $d+1$ we may move this false neighbor out of the neighborhood of $\mathbf{y}(k)$. By looking at every data point $\mathbf{y}(k)$ and asking at what dimension we remove all false neighbors, we will sequentially remove intersections of orbits of lower and lower dimension until at last we remove point intersections. At that juncture we will have identified that d_E where the attractor is unfolded [KBA92].

We now need a criterion for when a point $\mathbf{y}(k)$ and its nearest neighbor $\mathbf{y}^{NN}(k)$ as seen in dimension d are near or far in dimension $d+1$. In going from dimension d to $d+1$ the additional component of the vector $\mathbf{y}(k)$ is just $s(k+dT)$, and the additional component of the vector $\mathbf{y}^{NN}(k)$ is $s^{NN}(k+dT)$, both of which we know. Comparing the distance between the vectors $\mathbf{y}(k)$ and $\mathbf{y}^{NN}(k)$ in dimension d with the distance between the same vectors in dimension $d+1$, we can easily establish which are true neighbors and which false. We need only compare $|s(k+dT)-s^{NN}(k+dT)|$ with the Euclidian distance $|\mathbf{y}(k) - \mathbf{y}^{NN}(k)|$ between nearest neighbors in dimension d. If the additional distance is large compared to the distance in dimension d between nearest neighbors, we have a false neighbor. If it is not large, we have a true neighbor.

The square of the Euclidian distance between the nearest neighbor points as seen in dimension d is

$$R_d(k)^2 = \sum_{m=1}^{d} [s(k+(m-1)T) - s^{NN}(k+(m-1)T)]^2, \qquad (4.3)$$

while in dimension $d+1$ it is

$$\begin{aligned} R_{d+1}(k)^2 &= \sum_{m=1}^{d+1} [s(k+(m-1)T) - s^{NN}(k+(m-1)T)]^2, \\ &= R_d(k)^2 + |s(k+dT) - s^{NN}(k+dT)|^2. \qquad (4.4) \end{aligned}$$

The distance between points when seen in dimension $d+1$ relative to the distance in dimension d is

$$\sqrt{\frac{R_{d+1}^2(k) - R_d(k)^2}{R_d(k)^2}} = \frac{|s(k+dT) - s^{NN}(k+dT)|}{R_d(k)}. \qquad (4.5)$$

When this quantity is larger than some threshold, we have a false neighbor. As it happens the determination of which is true and which is false is quite insensitive to the threshold we use for changes once the number of data is enough to nicely populate the attractor. The search for nearest neighbors is made feasible when we have large amounts of data by using a *kd-tree* search scheme [FBF77, Spr91]. Establishing the neighbor relationships among N points using a kd-tree takes order $N(\log N)$ operations.

In practice for a large variety of systems we have examined, the threshold for

$$\frac{|s(k+dT) - s^{NN}(k+dT)|}{R_d(k)}, \qquad (4.6)$$

to define a false neighbor is a number about 15 or so. This varies with the number of data points for small data sets, but as soon as one has adequately sampled all regions of the attractor, the variation of false neighbors with the number of data is very small.

There is a subtler issue associated with identification of high dimensional signals when the number of data is limited. As we go to higher dimensions the volume available for data is concentrated at the periphery of the space since volume grows as distance to the power of dimension. This means that high dimensional signals will crowd to the edges of the space and no near neighbors will be close neighbors. To account for this we add to the previous criterion for a false neighbor the requirement that the distance added in going up one dimension not be larger than the nominal "diameter" of the attractor. This means that if

$$\frac{|s(k+dT) - s^{NN}(k+dT)|}{R_A} \tag{4.7}$$

is greater than a number of order two, the points $\mathbf{y}(n)$ and $\mathbf{y}^{NN}(n)$ would be labeled as false neighbors. Here R_A is the nominal "radius" of the attractor defined as the RMS value of the data about its mean. The results on determining d_E are rather insensitive both to the precise definition of R_A and to the value of the threshold as long as the number of data is not too small. Failure of either of these criteria signals that the distance change in going from dimension d to dimension $d + 1$ is too large and the neighbors will be declared false. Within substantial limits on the variation of the thresholds used one will see variations of the actual number or percentage of false nearest neighbors in dimensions less than d_E, but when zero false neighbors is reached, the number stays zero.

If we are presented with clean data from a chaotic system, we expect that the percentage of false nearest neighbors will drop from nearly 100% in dimension one to strictly zero when d_E is reached. Further, it will remain zero from then on, since once the attractor is unfolded, it is unfolded. If the signal is contaminated, however, it may be that the contamination will so dominate the signal of interest that we see instead the dimension required to unfold the contamination. If the contamination is very high dimensional, as we would anticipate from "noise", then we may not see the percentage of false nearest neighbors drop anywhere near zero in any dimension where we have enough data to examine this question. In the case of the Lorenz model where we can artificially control the percentage of contamination we add to the pure Lorenz chaos we will show how the determination of d_E degrades with noise.

This issue of contamination of data brings up the matter of what one considers "noise". Clearly using other tools such as Fourier analysis, chaos itself would have been classed as "noise", but it should be clear that this would be misleading. From the point of view of the dimensionality of the reconstructed phase space required to unfold the dynamical attractor "noise" is properly considered as just another dynamical system with a very high d_E. We have been calling it "noise" in quotes to emphasize that there is nothing special about this kind of contamination compared, say, to contamination of a five-dimensional signal by the addition of another chaotic but seventy-dimensional signal. Signal to contamination ratios make all the difference, as usual, and the determination of this ratio should be done in our time domain phase space. To repeat we determine d_E by looking at every data point in d-dimensional time delay reconstructed state space, namely, the vectors $\mathbf{y}(n)$. Then we ask whether the nearest neighbor $\mathbf{y}^{NN}(n)$ is true or false when we move to dimension $d + 1$. By looking at the nearest neighbor of every data vector we smartly step around any questions of how large a neighborhood should be and interview all data on an equal footing whether they appear in regions which are densely or thinly populated.

4.2.1   Lorenz Model

Our first example is constructed from $x(n) = x(t_0 + n\tau_s)$; $\tau_s = 0.01$ data from the Lorenz model. As determined above we use $T = 10$ to construct time delay vectors. In Figure 4.1 we display the percentage of global false nearest neighbors for this Lorenz data as a function of dimension. The percentage of false nearest neighbors goes to zero at $d_E = 3$, and as expected

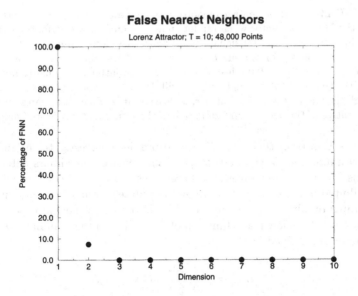

FIGURE 4.1.   The percentage of global false nearest neighbors for the dynamical variable $x(t)$ of the Lorenz model [Eq. (3.5)], using $T = 10$ as determined by average mutual information. The percentage of global false nearest neighbors drops to zero at $d_E = 3$ indicating that time delay phase space reconstruction can be accomplished with $x(t)$ data and its time lags: $[x(t), x(t + T\tau_s), x(t + 2T\tau_s)]$. The embedding theorem [Man81, Tak81, ER85, CSY91] would only assure us that $d_E = 5$ can be used as $d_A = 2.06$ for this system. $d_E = 3$ is the necessary integer dimension for unfolding the attractor from $x(t)$ data.

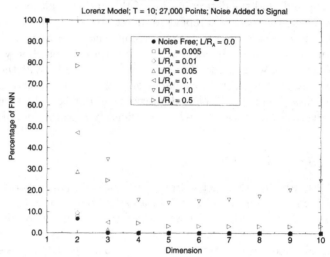

FIGURE 4.2.   The percentage of global false nearest neighbors for the $x(t)$ signal from the Lorenz attractor when random numbers uniformly distributed in $[-L, L]$ are added to the clean $x(t)$ signal. In the figure the global false nearest neighbors percentage is given for various L as a fraction of the RMS size of the attractor R_A. L/R_A ranges from 0.0 (clean data) to 1.0. The latter is a signal to noise ratio of 0 dB. It is clear that enough "noise" will defeat the false nearest neighbors test though the failure is graceful. Further, we learn that residual false nearest neighbors is indicative of the level of contamination of the signal that one is studying.

remains zero from there on. We are also able to investigate the influence of contamination on the percentage of false nearest neighbors determined in this fashion. To $x(n)$ from the Lorenz system we add a signal uniformly distributed in the interval $[-L, L]$ and populated from a "random" number generator on almost any computer. The false nearest neighbor percentage in Figure 4.2 is shown as a function of L relative to the nominal size of the attractor R_A defined by

$$R_A = \frac{1}{N} \sum_{k=1}^{N} |s(k) - s_{av}|,$$

$$s_{av} = \frac{1}{N} \sum_{k=1}^{N} s(k). \tag{4.8}$$

The ratio $\frac{L}{R_A}$ is a measure of the "noise to signal" in the total signal. We do not measure signal to noise in Fourier space since all the methods we use are in time domain in time delay phase space. From Figure 4.2 we see clearly that as the percentage of higher dimensional signal, the "noise",

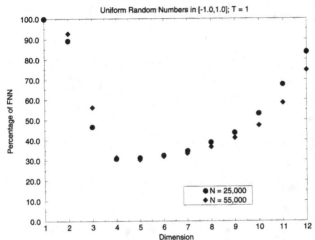

FIGURE 4.3.   The percentage of global false nearest neighbors for random numbers uniformly distributed in $[-1.0, 1.0]$ using $N = 25,000$ points (Solid Circles) and $N = 55,000$ points (Solid Diamonds). A time delay $T = 1$ is used. The average mutual information for "noise" or undersampled data may show no minimum, so a time delay of unity is suggested.

increases, the ability of the false nearest neighbor method degrades slowly and gracefully, until at $\frac{L}{R_A}$ of order unity the method indicates that we have a high dimensional signal, which, after all, is correct.

It is important that for large $\frac{L}{R_A}$ the fraction of false nearest neighbors does not fall to zero as d increases. This implies that the false nearest neighbors statistic provides us with a qualitative way to establish that the signal we see has residual 'noise' in it, and thus a way to distinguish 'noise' from low dimensional chaos. Just for comparison with the purposefully contaminated Lorenz system data, we display in Figure 4.3 the false neighbor plot for $\frac{L}{R_A} \to \infty$, that is just output from our random number generator with uniformly distributed noise. The false nearest neighbor calculation for output from a random number generator depends in detail on the number of data used. The larger the number, the sooner the percentage of false nearest neighbors rises to nearly 100%. We illustrate this dependence in Figure 4.3 where we show the false nearest neighbor percentage from the same random number generator for $N = 25,000$ and $N = 55,000$.

4.2.2   Nonlinear Circuit with Hysteresis

We return to the two output voltages from the NRL hysteretic circuit discussed earlier. Each voltage has an average mutual information function

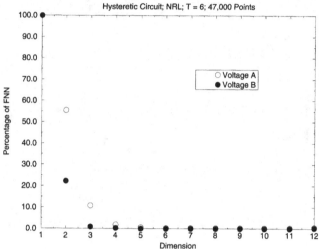

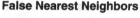

FIGURE 4.4.   The percentage of global false nearest neighbors for the two voltages $V_A(t)$ and $V_B(t)$ for the nonlinear hysteretic circuit. If we use data from $V_A(t)$ for time delay reconstruction of the attractor, we must use $d_E = 5$ to unfold the geometric figure of the attractor. When using $V_B(t)$ data and its time delays, $d_E = 3$ will work. This difference points out that not all dynamical variables and their time delays provide the same phase space reconstruction. They yield different coordinate systems when used with their time delays, and these may twist and fold the attractor sufficiently that different global embedding dimensions d_E are required. The two voltages must reveal the same local dimension d_L, and we will see that this is so.

with first minimum at $T = 6$. Now using this value of T we construct false nearest neighbors for each voltage data set. For the voltage $V_A(t)$ an embedding dimension of $d_E = 5$ is seen in Figure 4.4. For data from $V_B(t)$ we find in the same figure a global embedding dimension of $d_E = 3$. That these are different iterates an important point about time delay embedding mentioned above: **not all global reconstructed coordinate systems are the same even for different measurements from the same source.** In one coordinate system the attractor may be quite bent and folded and require more spatial coordinates to unfold than in another coordinate system. The maximum number is always given by the embedding theorem as the integer greater than $2d_A$, but the number actually required by the data may be less for some choice of coordinates. Indeed, if we were to somehow discover the true set of dynamical variables, it seems clear that just the integer greater than d_A would be required.

We will note later but state now that if all one wishes to do is determine the fractal dimension, then using the integer dimension just greater than d_A works [DGOS93]. This is quite a different requirement than separating orbits from each other when they are artificially nearby due to projection. The goal in our case is to establish a global space in which deterministic features of the data are retained, so points $\mathbf{y}(n)$ follow each other in time sequence if the points in the original, but unknown, dynamical source follow each other in time sequence. A dimension which works for determining d_A may fail when one wants to build predictive models. Indeed, in the present example from the hysteretic circuit we see that with either V_A or V_B one may determine d_A using a three-dimensional embedding. If one wishes to have a deterministic evolution in time delayed V_A space, using dimension five is required.

4.3   A Few Remarks About Global False Nearest Neighbors

The basic geometric idea in the embedding theorem is that we have achieved an acceptable unfolding of the attractor from its values as seen projected on the observation axis when the orbits composing the attractor are no longer crossing one another in the reconstructed phase space. The algorithmic implementation of this has proven a valuable tool in establishing a necessary global dimension for unfolding chaotic dynamics. The false nearest neighbors calculation is simple and fast. As we have seen, it is also graceful in its failure as one increases contamination of the signal of interest. An interesting and useful byproduct of the false nearest neighbors calculation is an indication of the noise level in a signal. Indeed, in one of our examples near the end of this book we will use this piece of information to point to the role of intrinsic quantum noise in the operation of a solid state laser. Without false neighbors to guide us we might have been unable to distinguish "noisy looking" laser intensity in the regime where it is just chaotic from the regime in which quantum effects are important.

The results on d_E from the two measurements from the chaotic circuit remind us that the global embedding dimension is tied to the coordinates out of which one chooses to construct the state space. It is almost obvious, especially in retrospect, that one need not find the same global d_E for an arbitrary nonlinear change of coordinates. The minimum d_E presumably belongs to the "correct" original coordinates in which the system evolves. Finding the nonlinear transformation from time delay coordinates to that best set of coordinates has eluded any attempts I have seen or tried. From a practical point of view, perhaps it is not a useful quest as one can answer most physical questions about the source of the observations in whatever coordinate system time delay embedding provides.

The dimension of the dynamics with which one works, or better which is being sensed by one's instruments, may well depend on the resolution of those instruments. This is quite clear from the results on boundary layer chaos which we shall discuss in the Chapter 10. There an instrument 70 times larger than the identified "optimal" size for resolving the fluid dynamical activity on all spatial scales was used for the measurements. The filtering by the instrument resulted in a much smaller number of dynamical variables being sensed during the observations.

The message is more or less this: if one filters out some characteristic part of a total signal so that the amplitude of the measured signal comes, say, 99.5 percent from the unfiltered part of the signal, that doesn't mean there is no dynamics associated with the filtered out part of the signal, just that we have reduced its amplitude to a numerical level where its presence is unimportant in the remaining time series. Imagine adding to a sine wave $A\sin(t)$ a signal from the Lorenz system with amplitude $10^{-7}A$. We will see only a sine wave through all our tests: $I(T)$ will reflect the regular, periodic structure, and false nearest neighbors will instruct us to use $d_E = 2$.

This process of filtering is quite familiar throughout physics. When considering the dynamical equations for the hysteretic circuit used as an example in this book we would not be inclined at all to begin with Maxwell's equations and then smooth or filter the detailed electric field fluctuations in one of the resistors. We are quite comfortable with assigning the lumped parameter of a resistance R to that circuit element and then using Ohm's law in its crude, but useful form with the voltage drop across the element being just the current times R. It is only when we focus on capturing every possible degree of freedom in a dynamical system and concentrate rather heavily on the very interesting idea from the embedding theorem that every dynamical variable is coupled into all others and their effect can **in principle** be revealed by studying any single variable, that we open ourselves up to expecting to see all dynamics through every instrument. Clearly that is not going to be the case and the estimate of d_E using false nearest neighbors or whatever other method one chooses is simply not going to reveal dynamical variation which is suppressed by physical or other filtering.

This is good news and bad news. The **good news** is that filtering out spatial variations on small scales, perhaps even associated with high frequencies as well, can lead to a practical system of lower dynamical dimensions. Later we shall examine the time series of the volume of the Great Salt Lake and show it is captured in $d_E = 4$. There is no real doubt in anyone's mind that the dimension of the active degrees of freedom in weather or climate exceeds four. Indeed, using data from large numerical simulations of climate, one finds high dimensional systems, and even very intermittent dynamics for variables such as local precipitation. The Great Salt Lake volume averages over regions nearly 300 km on a side, and in this way simplifies the critical dynamics. This means one has a chance to describe in a detailed, quantitative fashion, regionally averaged climate variations even when the spatially local dynamics may be quite complex and high dimensional. A similar lesson is learned with the boundary layer chaos.

By spatially filtering out the small scale fluctuations, one has identified a dynamical system with which one can work and thus accomplish further understanding of the boundary layer flows.

The **bad news** is that one must be alert to the possibility that interesting dynamical aspects of a problem have been filtered out by one's instruments. This then requires a closer look at how experiments are done than might be called for by a review of the actions of instruments in Fourier domain. Simply noting that a signal has been filtered may not tell what effect this has on the apparent dynamics. Filtering a signal with a moving average filter which replaces the original measurements with linear combinations of the measurements, in principle is a harmless renaming of the dynamical variables. Filtering with an autoregressive filter [ABST93] can change the apparent dimension of the system. It is not enough to ask what Fourier components or other linear transform of one-dimensional data components have been retained to establish what is happening dynamically in multidimensional space. The real activity of nonlinear systems in chaotic motion is revealed in the multivariate space which is established globally by false nearest neighbors or equivalent techniques.

4.4 False Strands

A cautionary note about the use of global false nearest neighbors is in order. If the data one has is from a set of differential equations, then **oversampling** the data can lead one astray. The first signal one may have of oversampling is that the first minimum of average mutual information comes at a value of T which is a few tens of the sampling time τ_s. If one has oversampled, then it is possible that the nearest neighbor whose trueness or falseness one is examining is just the point on the trajectory displaced by $\pm\tau_s$. This is always a true neighbor and the statistics of what false neighbors have been projected into a neighborhood can be biased. As the number of data increases this problem will fade away, but for realistic data sets, one may be able to do little about increasing the sample size or increasing the sampling time while maintaining a large data set.

The simplest repair of this problem is to systematically remove data points thus effectively increasing τ_s by eliminating every other data point and then every other one again until T is a small integer times the effective sampling time. This has the disadvantage of throwing away hard won data. One may instead systematically eliminate from consideration in testing for false neighbors any neighbor which has a time index $\pm K\tau_s$ where K is some small integer. This also works rather well and focuses attention on neighbors with quite distant time indices which would come either from parts of the orbit folded back to the neighborhood in question or from points falsely projected into the neighborhood by working in too small a dimension. These are the points one wishes to distinguish in determining d_E. A more powerful technique, requiring a bit more computation, is to identify whole strands of data points which are in each other's neighborhoods and then ask what happens to the average distance between points on these strand pairs [KA95]. This effectively eliminates the problem of oversampling by carefully making sure the sets of points whose neighborliness is being examined correspond to pieces of orbit which might have falsely been projected into each other's region of phase space.

4.5   Other Methods for Identifying d_E

Since the introduction of the time delay embedding procedure [PCFS80, ER85], there have been more than a few methods for determining d_E. One popular one has been to look for that dimension at which the correlation integral [GP83], which we will discuss in Chapter 5, becomes independent of the dimension of the reconstructed space. Ding, et al [DGOS93] showed that this independence typically comes at the integer dimension $d > d_A$ and does not always unfold the attractor. This should serve as a warning that the dimension in which one can compute some interesting quantity is not necessarily the same as the dimension in which the predictability of the underlying deterministic source is achieved. It is important to carefully distinguish what it is one wishes to achieve by evaluating some quantity associated with chaotic motion and then evaluate just that quantity which achieves one's goal.

Another direct method is given by the "true vector field" approach of Kaplan and Glass [KG92]. They note that when the attractor is unfolded, the vector field associated with $x(t) \rightarrow x(t + 1)$ dynamics, namely, the vector function $F(x)$ in $x \rightarrow F(x)$, is unambiguous. This means that if we can construct a local representative for this vector field, then the direction of all flow vectors in any neighborhood will point in the same direction when the attractor is unfolded. If we are in a dimension which is too low to unfold the attractor, then in some neighborhoods two or more directions of the vector field will be folded on top of each other and the direction of the vector field in those neighborhoods will not be unique. This algorithm also goes to the heart of the embedding process: one is looking for a space in which the direction of evolution in each neighborhood of the attractor is unambiguous. This is guaranteed by the uniqueness theorems on the solutions of the dynamics in the original $x(t)$ coordinates, and it is the goal of the embedding theorem to provide a substitute space where the sense of the evolution is preserved so prediction in that space can be made with the same lack of ambiguity as in the original space. If there is false crossing in a reconstructed space, we will not be able to tell by looking at a local neighborhood where the evolution is going, and prediction will be compromised.

The idea then is to take points $\mathbf{y}(n)$ in dimension d and form the unit directional vectors

$$\frac{\mathbf{y}(n+1) - \mathbf{y}(n)}{|\mathbf{y}(n+1) - \mathbf{y}(n)|}, \qquad (4.9)$$

at each location on the attractor. Then we define a neighborhood of a given size associated with each point $\mathbf{y}(n)$ on the attractor and determine whether all the unit directional vectors in that neighborhood point in the same direction. More precisely, Kaplan and Glass test whether the distribution of directions in a neighborhood is consistent with being random or whether it arises from a nonstatistical, deterministic rule. In this manner they establish at what dimension the vector fields so constructed are all true. While they have not done it point by point, they could well have looked at each point on the attractor $\mathbf{y}(n)$ and its nearest neighbor $\mathbf{y}^{NN}(n)$ and tested what percentage of the time the vectors $\mathbf{y}(n+1) - \mathbf{y}(n)$ and $\mathbf{y}^{NN}(n+1) - \mathbf{y}^{NN}(n)$ are parallel with some tolerance. This altered version of true vector fields would now look rather like false nearest neighbors and avoid the tricky question of what constitutes a neighborhood when data become sparse.

A numerical flaw with the true vector field algorithm can arise because to form the directional vectors one must take the difference between numbers each of which in many experimental situations may not be known very accurately. As ever, when taking the difference between inaccurately known numbers, one is in effect making a high pass filter numerically and this may emphasize the roundoff errors in the data and produce unwelcome contaminations. This curse of creeping contamination is also associated with the desire to determine accurate evaluations of the time derivatives of the observed quantities $s(t)$ to use as coordinates for the reconstructed data vector $\mathbf{y}(n)$ [PCFS80].

4.6   The Local or Dynamical Dimension d_L

Once one has determined the integer global dimension required to unfold the attractor on which the data resides, we still have the question of the number of dynamical degrees of freedom which are active in determining the evolution of the system as it moves around the attractor. As one example of the relevance of this question we can look back at the data from the chaotic hysteretic circuit of Section 5.2.2. One of the observed variables V_A gave a global embedding dimension $d_E = 5$, and another measured variable V_B indicated that $d_E = 3$ would work. One of these must be too large. The local dimension of the dynamics must be the same for every way we have of sampling the data even though our methods for reconstructing a phase space in which to globally unfold an attractor can quite naturally produce a space of dimension larger than the dynamics of the signal source.

Another example which needs no elaborate explanation is a system which has two incommensurate frequencies in its motion:

$$s(t) = A_1 \sin(2\pi f_1 t + \phi_1) + A_2 \sin(2\pi f_2 t), \qquad (4.10)$$

where the ratio $\frac{f_1}{f_2}$ is irrational, ϕ_1 is some phase, and the A_j are arbitrary amplitudes. The flow associated with this behavior lies on a two-torus, but to unfold the motion on this torus we must embed the system in $d_E = 3$. The number of dynamical variables is clearly two. We now describe a method which would tell us that two dynamical degrees of freedom are active in this flow and identifies the correct number of active degrees of freedom in other cases as well.

Active degrees of freedom is a somewhat intuitive concept. It is clear when we speak about ordinary differential equations or maps of low dimension, for then all the variables of the system will be involved in low dimensional behavior as revealed in observations. If we have a high dimensional system of ordinary differential equations or a partial differential equation with a mathematically infinite number of degrees of freedom, it is somewhat harder to be precise about the active degrees of freedom for that may in practice depend on the resolution of our instruments.

An operational definition may suffice to give us a clear physical sense of what we might mean by "active degrees of freedom". In a system of rather high dimension, there may be several positive Lyapunov exponents which are responsible for the apparent instability of an orbit throughout phase space. There may be a zero global Lyapunov exponent reflecting the fact that the system is described by differential equations labeled by a continuous time variable. There will also be many negative Lyapunov exponents assuring us that the system is dissipative and governing the decay of transients in the orbits. The number of numerically important negative exponents will provide a limit on the degrees of freedom which act in a numerically important enough fashion as to be visible in any given measurement. If one of the Lyapunov exponents is very negative, it will cause such rapid decay back to the relevant attractor that we may never be able to measure its influence on a trajectory, and it will appear "inactive" from any of the tests, false neighbors or otherwise, that we may apply to the data. Now admittedly this is a somewhat heuristic painting of the notion of active degrees of freedom, but it may suffice for purposes of giving guidance as to what we mean as we proceed to identify local or dynamical dimensions. There is a clear analogy here with the notion of "slaved variables" or variables which are frozen out by actions on different time scales in various processes.

In our discussion of chaotic laser fluctuations in Section 11.1 we will see such a decrease in dynamical degrees of freedom in an experimental example. Rather than the potentially large number of modes of light allowed in a laser cavity, we will show evidence using the methods discussed here that only a few (seven there) are operating. If the source of our observations is a partial differential equation, for example, the local dimension of the dynamics is infinity. This will not appear in any test we make if the data has finite resolution. So, finally, we may converge on the idea that local or dynamical dimension consists of those degrees of freedom $d_L \leq d_E$ which are numerically large enough to express themselves in our observations. If we see the source through a spatial or temporal filter which suppresses degrees of freedom in a substantial numerical way, well then, we will have no choice but to conclude pragmatically that those degrees of freedom are absent for any further purposes of describing or utilizing what we have learned about the source of our measurements from those filtered data. Higher resolution experiments may reveal those "lost" degrees of freedom, but that is for further observations to uncover. It is quite useful to recall that the various analyses described in this book are directed toward constructing prediction or control models for the source of the observations or perhaps directed toward classifying the source by dynamical invariants (as in the next Chapter). The determination of d_L can be operational rather than precise as in the case of fluid dynamics or other sources governed by partial differential equations. The experimental resolution or the natural filtering of the data may be critical in determining a value of d_L with which one can make high quality predictions. If one could have higher resolution data, then perhaps d_L would be larger, but until that data is available, one cannot know.

We will discuss two basic ideas for distinguishing the $d_L \leq d_E$ true or active or dynamical degrees of freedom from those dimensions d_E required to unfold the attractor:

i **compute local and global Lyapunov exponents forward and backward in time** in dimension d_E. True exponents, of which there will be $d_L \leq d_E$, will reverse under this operation. The false exponents, of which there will be $d_E - d_L$, will do something else [ER85, Par92, AS93]. The local and global Lyapunov exponents will be explained below. This method works on good, clean data. It is very susceptible to the presence of contamination.

ii **evaluate the percentage of local false nearest neighbors.** We will explain this in some detail in this section [AK93]. This turns out to be quite robust against contamination.

4.7   Forward and Backward Lyapunov Exponents

The first method compares local Lyapunov exponents, properly defined in the next chapter, evaluated forward in time along the data set and then

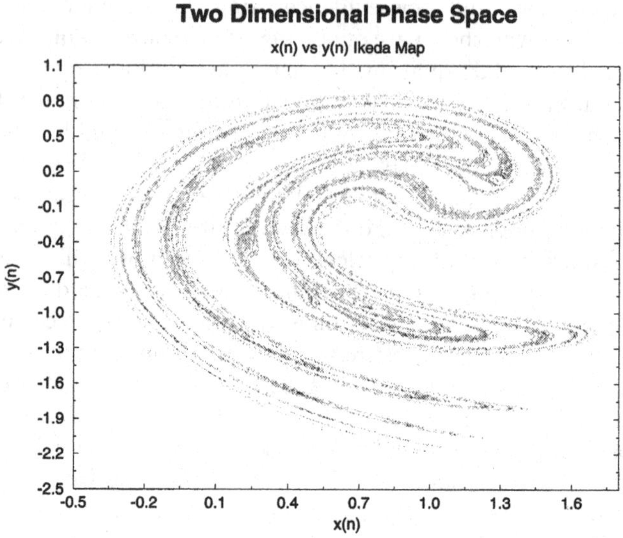

Two Dimensional Phase Space

x(n) vs y(n) Ikeda Map

FIGURE 4.5.   The phase space portrait of the Ikeda map, Equation (4.11), using the dynamical variables $(x(n), y(n))$ directly from the equations.

evaluated backward in time along the same data set. As we suggested in the qualitative discussion in Section 1.4, exponential growth in time as $e^{t\lambda}$ is generic in unstable nonlinear systems. If we switch the sign of time, then this is equivalent to switching the sign of λ, and that is the idea behind the method we discuss now.

The idea is illustrated by the example of the Ikeda map [Ike79, HJM85] which arises in the analysis of the passage of a pumped laser beam around a lossy ring cavity. The amplitude and phase of the beam are summarized in the complex number $z(n) = x(n) + iy(n)$ which evolves as

$$z(n+1) = p + Bz(n) \exp\left[i\kappa - i\frac{\alpha}{1 + |z(n)|^2}\right], \tag{4.11}$$

where the parameter values are $p = 1.0$, $B = 0.9$, $\kappa = 0.4$, and $\alpha = 6.0$.

The attractor is shown in Figure 4.5 in original coordinates $(x(n), y(n))$. An attempt to reconstruct the attractor using the time delay coordinates $[x(n), x(n+1)]$ is illustrated in Figure 4.6. Clearly $d_E = 2$ will not work for this $x(n)$ time series. The global false nearest neighbors method indicates $d_E = 4$ without ambiguity. This is shown in Figure 4.7.

Using the methods we discuss in some detail in the next chapter for establishing the Lyapunov exponents of a dynamical system from the observation of a single scalar data stream, we then computed the local Lyapunov exponents $\lambda_i(\mathbf{x}, L); i = 1, 2, \ldots, d_L$. These local exponents tell us how per-

Two Dimensional Phase Space

x(n) vs x(n+1) Ikeda Map

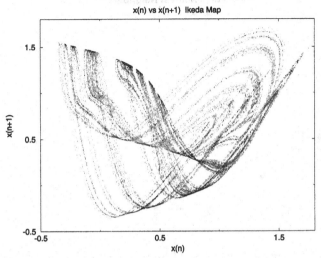

FIGURE 4.6.   The reconstructed phase space portrait of the Ikeda map, Equation (4.11), using the dynamical variables $(x(n), x(n+1))$. it is clear to the eye that the attractor is not unfolded in these variables even though the system is a two-dimensional map of the plane to itself. We expect that using false nearest neighbors we will identify another integer $d_E > 2$ which unfolds the Ikeda attractor using time delay reconstruction.

turbations at some location $\mathbf{x}$ on the attractor to the orbit behave as we follow the perturbation for L steps along its evolution. In d_E dimensions one evaluates d_E Lyapunov exponents. Some of these are sure to be false, if the dimension of the dynamics is less than d_E, as is the case for the Ikeda map. Evaluating the $\lambda_a(\mathbf{x}, L)$ backward in time reverses the true exponents, and comparing the forward and backward evaluations for various values of the local dimension chosen for the determination of the number of $\lambda_a(\mathbf{x}, L)$, tells us the local dynamical dimension.

For the Ikeda map we used 500 starting locations $\mathbf{x}$ on the attractor and followed the orbit for $L = 2^k; k = 1, 2, \ldots, 15$ steps from each initial $\mathbf{x}$. The average for each $\lambda_a(\mathbf{x}, L)$ at fixed L over the 500 initial $\mathbf{x}$ locations was then made for exponents computed forward along the data and then backwards along the same data. A local neighborhood to neighborhood polynomial fit to the data [BBA91] was made with a third order polynomial in $d_L \leq d_E$. All required distances were evaluated in $d_E = 4$.

We report the average local Lyapunov exponent $\bar{\lambda}_a(L)$ as a function of L. In Figure 4.8 we display the results of the calculation for $d_E = 4$ and $d_L = 4$. Forward exponents are indicated with open symbols and minus the backward exponents are indicated with solid symbols. Only two of the four

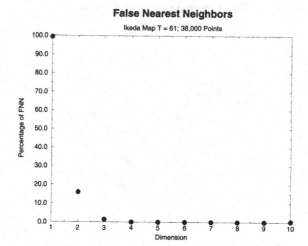

FIGURE 4.7.   The percentage of global false nearest neighbors for the Ikeda map, Equation (4.11) using $T = 1$ time delays of the dynamical variable $x(n)$. It is clear that the global embedding dimension $d_E = 4$ is chosen for unfolding the Ikeda attractor, Figure 4.5, in the coordinate system of time delays of $x(n)$. This is also the sufficient dimension as the box counting dimension of the Ikeda attractor is $d_A \approx 1.8$, and the criterion for a sufficient dimension $d_E > 2d_A$ yields $d_E = 4$.

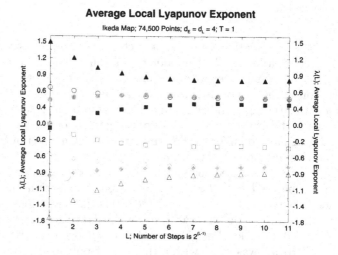

FIGURE 4.8.   The average local Lyapunov exponents $\bar{\lambda}_a(L)$ evaluated from Ikeda map, Equation (4.11), $x(n)$ data using $d_E = d_L = 4$. The average exponents are evaluated forward and backward in time [Par92, AS93], and we can see that only two of them agree indicating that $d_L = 2$ even though we must use $d_E = 4$ to unfold the attractor in time delay embedding coordinates.

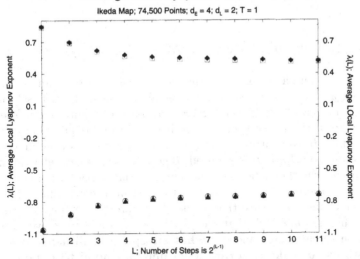

Average Local Lyapunov Exponent

Ikeda Map; 74,500 Points; d_E = 4; d_L = 2; T = 1

FIGURE 4.9.   The average local Lyapunov exponents $\bar{\lambda}_a(L)$ evaluated from Ikeda map, Equation (4.11), $x(n)$ data using $d_E = 4$, $d_L = 2$. The average exponents are evaluated forward and backward in time [Par92, AS93], and we can see that both agree supporting the conclusion from Figure 4.8 that $d_L = 2$.

exponents are the same in the forward and (minus) backward computations. In Figure 4.9 $d_E = 4$ as usual, but $d_L = 2$ is established. Now the correspondence of forward and (minus) backward exponents is quite striking. At $L = 2^{10}$ we have for the forward exponents 0.512 and −0.727 and for the backward exponents −0.510 and 0.731.

We performed the same computations in $d_E = 5$ to test the efficacy of the method when one has chosen an incorrect embedding dimension. Essentially the same pattern emerges, though the inability to clearly determine false exponents until d_L is correct is magnified. At $L = 2^{12}$, for example, the forward exponents for $d_L = 3$ are 0.545, −0.655, and −0.844, while the backward exponents are 0.758, −0.479, and −0.783. At $d_E = 5$; $d_L = 2$. and $L = 1000$ the forward exponents are 0.530 and −0.753 while the backward exponents are 0.753 and −0.541.

For purposes of illustration we have chosen here an especially nice example where the discrimination of d_L by forward and backward Lyapunov exponents works neatly and cleanly. This, unfortunately, is not always the case. Errors in the data or noise added to the data by instruments, computer roundoff, or a communications channel make the determination of Lyapunov exponents rather difficult at times. Further a nice clean reversal of true exponents in time is simply not always seen. Tracking the average local exponents when it can be done cleanly is a sure way to identify d_L, but *caveat emptor*.

4.8   Local False Neighbors

Just as we were able to unfold the global phase space for an attractor by looking at the presence of nearest neighbors which arrive falsely in each others' neighborhood by projection, so may we inquire about the **local** structure of the phase space to see if **locally** one requires fewer dimensions than d_E to capture the evolution of the orbits as they move on the attractor. To achieve this we need to go beyond the geometric arguments which underlie the embedding theorem and the global false nearest neighbors construction. Our approach is to work in a global dimension which is large enough to assure that all neighbors are true, namely, some working dimension d_W such that $d_W \geq d_E$. In this space we move to a point $\mathbf{y}(k)$ on the attractor and ask **what subspace of dimension $d_L \leq d_E$ allows us to make accurate local neighborhood to neighborhood maps of the data on the attractor.** To answer this question we must define a neighborhood which we do by specifying the number of neighbors N_B of the point $\mathbf{y}(k)$ and then provide a local rule for how these points evolve in one time step into the same N_B points near $\mathbf{y}(k+1)$. We then test the quality of these predictions and seek a value of d_L where the quality of the predictions become independent of d_L and of the number of neighbors N_B.

How shall one choose the coordinates for this local dimension d_L from among the many possible choices? Our choice is based on the idea of a local principal component analysis [GV89] of the data. This locally selects those directions in d_W-dimensional space which contain the majority of the data in a least-squares sense. We make this local principal component decomposition in dimension d_W by forming the sample covariance matrix among the N_B neighbors $\mathbf{y}^{(r)}(k); r = 1, 2, 3, \ldots, N_B$ of $\mathbf{y}(k)$. The sample covariance matrix is the $d_W \times d_W$ quantity

$$\mathbf{R}(k) = \frac{1}{N_B} \sum_{r=1}^{N_B} [\mathbf{y}^{(r)}(k) - \mathbf{y}_{av}(k)][\mathbf{y}^{(r)}(k) - \mathbf{y}_{av}(k)]^T, \tag{4.12}$$

where

$$\mathbf{y}_{av}(k) = \frac{1}{N_B} \sum_{r=1}^{N_B} \mathbf{y}^{(r)}(k). \tag{4.13}$$

The eigenvalues of this covariance matrix are ordered by size, and as our basis for the local d_L dimensional space we choose the eigendirections associated with largest d_L eigenvalues. This choice guarantees that "most" of the data, in the sense of capturing the variation of the data about its local mean, is along the directions we have selected. The final answer we arrive at for the best local or dynamical dimension d_L should be independent of this particular way of choosing local coordinates, but this choice turns out to be robust against noise. We do not have a theorem or even statistical proof that choosing the d_L largest eigendirections of the sample covariance matrix is a good selection of $d_L < d_W$ directions for examining local properties. It is a perfectly safe and workable choice, and it is certainly not optimized in any mathematical fashion. Were one to choose another selection of d_L coordinates, that would be perfectly fine as long as one avoids accidental projection of the vectors in d_W space onto a subspace where there is no data. This would result in singular coordinates. The convenient selection of the dominant principal components is a good way to avoid this problem.

Once we have our local basis of d_L d_W-dimensional vectors, we form the projection of the d_W-dimensional data vectors $\mathbf{y}^{(r)}(k)$ onto these d_L eigendirections. Call these vectors $\mathbf{z}^{(r)}(k)$. These constitute our N_B local d_L dimensional vectors at "time" k. Now we find the vectors $\mathbf{z}^{(r)}(k; \Delta)$ which evolve from the $\mathbf{z}^{(r)}(k)$ in Δ time steps.

For this construct a local polynomial map

$$
\begin{aligned}
\mathbf{z}^{(r)}(k) &\rightarrow \mathbf{z}^{(r)}(k; \Delta), \\
\mathbf{z}^{(r)}(k; \Delta) &= \mathbf{A} + \mathbf{B}\mathbf{z}^{(r)}(k) + \mathbf{C}\mathbf{z}^{(r)}(k)\,\mathbf{z}^{(r)}(k) + \dots, \quad (4.14)
\end{aligned}
$$

which takes the vectors $\mathbf{z}^{(r)}(k)$ into their counterparts $\mathbf{z}^{(r)}(k; \Delta)$. The coefficients $\mathbf{A}, \mathbf{B}$ and $\mathbf{C}, \dots$ are determined in a least-squares sense. This means we make a polynomial from the components of the $\mathbf{z}^{(r)}(k)$ and minimize

$$
\sum_{r=1}^{N_B} |\mathbf{z}^{(r)}(k; \Delta) - \mathbf{A} - \mathbf{B}\mathbf{z}^{(r)}(k) - \mathbf{C}\mathbf{z}^{(r)}(k)\,\mathbf{z}^{(r)}(k) - \dots|^2, \quad (4.15)
$$

where $\mathbf{A}, \mathbf{B}, \mathbf{C}, \dots$ are constant tensors of appropriate rank. Typically we need only local linear maps since we are working locally. If we have a lot of data, local linear will work just as well as local quadratic or higher order local maps, and local linear may well suffice. If the data becomes sparse, the neighborhood size required to contain enough points N_B to allow accurate determination of the coefficients in the local map could grow to be unacceptably large. In this case local linear maps may work best simply because they put smaller demands on the available data. If local linear maps are to be used, we encourage the use of local quadratic maps as a check on the calculations. Reassurance of this sort is clearly somewhere between folklore and careful checking of calculations. In any case it may be important to be sure of the choice of d_L. In what follows we stick to local linear maps and provide the algorithm with sufficient data. When we have taken our own advice and checked the computations, no change has occurred in what we will report below.

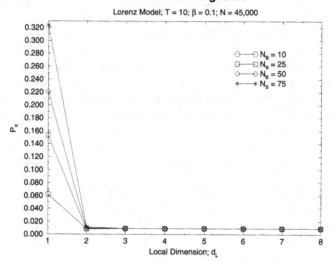

FIGURE 4.10.   Percentage of bad predictions P_K as a function of local dimension d_L and number of neighbors N_B for time delay vector data from the $x(t)$ dynamical variable from the Lorenz attractor, Equation (3.5). A time delay of $T = 10$ was used as determined by average mutual information. Forty five thousand data points were used, and the horizon of bad predictions was set at βR_A with $\beta = 0.1$. Changing β alters the number of local false neighbors but does not change the d_L where these false neighbors, or equivalently P_K, becomes independent of d_L and N_B. The choice of $N_B = 10, 25, 50$, and 75 is not critical, just a range to expose independence of N_B when it occurs.

Having determined the local linear map, we ask how well it predicts forward in time. In particular we ask when the local polynomial prediction map **fails** because it makes a prediction error in Δ steps which is some finite fraction of the size of the attractor

$$R_A = \frac{1}{N} \sum_{k=1}^{N} |s(k) - s_{av}(k)|, \qquad (4.16)$$

as defined above. When the percentage of bad predictions becomes independent of d_L and is also insensitive to the number of neighbors N_B, we may say that we have identified the correct local dimension for the active dynamical degrees of freedom.

This local criterion for the dynamical degrees of freedom mixes both the geometric idea of false nearest neighbors and the dynamical idea of quality of prediction. The latter plays a role when we make models later, but its appearance here comes because we have no fully geometric method for identifying how unfolded an attractor is locally. So we have added the

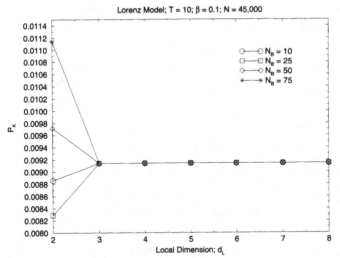

FIGURE 4.11.   An expanded version of Figure 4.10 showing how the P_K selects $d_L = 3$ in a clear fashion for the Lorenz system. The dimension d_A of the Lorenz attractor is nearly two, so it is not surprising that near $d_L = 2$ the P_K has come very close to independence of d_L and N_B.

dynamical criterion that whatever local dimension we choose had better be adapted to the dynamics as well as the geometry. This clearly goes beyond the Takens' embedding theorem, as it must, because that theorem makes statements only about **global** aspects of the attractor.

It cannot hurt to reemphasize that even though this method works very well in practice, we do not have a mathematical theorem to guide us here, so it may well be possible to devise a better way to identify the local dynamical dimension for a data set. Happily it is robust against as much as 5 to 10% contamination **at the scale of the attractor size** R_A which itself may be as large as 100 times the extent of a local neighborhood.

4.8.1 Lorenz Model; Ikeda Map

Using "observations" of $x(n)$ from the Lorenz model, we display in Figure 4.10 the percentage of bad predictions as a function of N_B and d_L. We used $d_W = 8$ in these calculations. The local false neighbors become nearly independent of both parameters near $d_L = 2$, and by $d_L = 3$, the dependence on N_B and d_L is gone. If we enlarge this picture as in Figure 4.11, we see that $d_L = 3$ is appropriate. Recalling that the fractal (box counting) dimension of the Lorenz attractor is $d_A \approx 2.06$, it is not surprising that we would have to look carefully at the choice between $d_L = 2$ and $d_L = 3$.

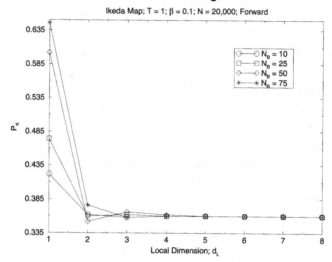

FIGURE 4.12.   Percentage of bad predictions P_K as a function of local dimension d_L and number of neighbors N_B for time delay vector data from the $x(n)$ dynamical variable from the Ikeda map, Equation (4.11). A time delay of $T = 1$ was used. Twenty thousand data points were used, and the horizon of bad predictions was set at βR_A with $\beta = 0.1$. These local false nearest neighbor evaluations were made reading the data forward in time.

This example is not as striking as the use of the method on data from the Ikeda map discussed earlier. We saw that global false nearest neighbors required $d_E = 4$ while the dynamics was clearly two dimensional. In Figure 4.12 we see the results of local false nearest neighbors evaluated on data from the real part of the Ikeda map dynamical variable, and in Figure 4.13 we see the same for local false neighbors evaluated backward on the same data set. The clear signal that $d_L = 2$ needs no further comment. What is more striking is that, in contrast to the method of backward and forward local Lyapunov exponents, this algorithm is rather robust against contamination by other signals. In Figure 4.14 we see how the addition of the output of a high dimensional random number program affects our ability to select $d_L = 2$ for the Ikeda map.

In Figure 4.15 we display the local false nearest neighbor curves for the data $x(n)$ from the Lorenz model with various levels of contamination as a function of d_L and N_B. These are noise levels of uniformly distributed random numbers in the interval $[-L, L]$ and the percentages of contamination are quoted as $\frac{L}{R_A}$, with R_A as above. We see that the local false nearest neighbor test fails gracefully as noise is added to the signal. Eventually we

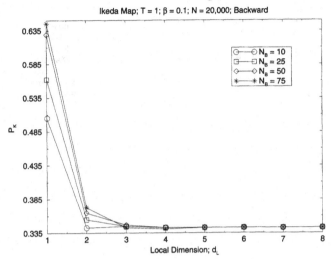

FIGURE 4.13.   Percentage of bad predictions P_K as a function of local dimension d_L and number of neighbors N_B for time delay vector data from the $x(n)$ dynamical variable from the Ikeda map, Equation (4.11). A time delay of $T = 1$ was used. Twenty thousand data points were used, and the horizon of bad predictions was set at βR_A with $\beta = 0.1$. These local false nearest neighbor evaluations were made reading the data backward in time. Combined with the computations shown in Figure 4.12 we can conclude that $d_L = 2$ is likely for the Ikeda map.

learn only that there is a lot of noise, that is, high dimensional stuff, and that is precisely what the local false nearest neighbors should tell us.

4.8.2   Nonlinear Circuit with Hysteresis

In the example of the hysteretic nonlinear circuit we are in the fortunate position of having measurements from two independent voltages. Recall from Section 5.2.2 that for voltage V_A from this circuit, time delay embedding required $d_E = 5$ to globally unfold the attractor. At the same time measurements of voltage V_B only required $d_E = 3$. If we now look at the local false nearest neighbors for V_A as displayed in Figure 4.16 we see that $d_L = 3$ is chosen which means that in the coordinates constructed from $V_A(n)$ and its time delays, we would use dimension five to unfold the attractor globally but only require local three-dimensional models for the dynamics. When we come to Lyapunov exponents below, we will see that this makes a substantial difference in the information we would extract from this system. Figure 4.17 contains the information on local false nearest neighbors from reconstructed time delay space made out of $V_B(n)$ and its

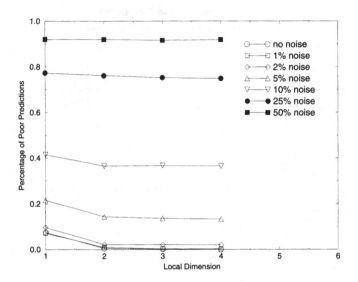

FIGURE 4.14.   Percentage of bad predictions P_K as a function of local dimension d_L and 'noise' level at $N_B = 40$ for time delay vector data from the $x(n)$ dynamical variable from the Ikeda map, Equation (4.11). A time delay of $T = 1$ was used. Twenty thousand data points were used, and the horizon of bad predictions was set at βR_A with $\beta = 0.5$. These local false nearest neighbor evaluations were made reading the data forward in time. Until the noise level reaches 10% measured **globally**, $d_L = 2$ is selected. This noise level is much larger when seen locally.

time delays. This confirms that $d_L = 3$ is here consistent with the $d_E = 3$ found for the $V_B(n)$ of coordinates.

   This is a useful lesson since one doesn't always have a choice of what measurements are available. However, if there is a choice, this strongly suggests one examine all available sets of data to establish which would provide the most economical global description of the dynamics. With several choices of observations available, one should check carefully using local false nearest neighbors that the local dimension for each is the same. This is a matter of self-consistency, of course, but could reveal problems with one or more data sets, Further it underlines the importance of not taking the results of a **global** embedding dimension and immediately concluding that the local dimension is that as well. This example and the numerical example mentioned before of the Ikeda map with $d_E = 4$ and $d_L = 2$, make this clear.

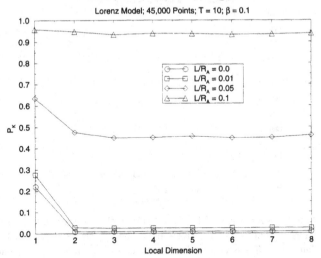

Local False Nearest Neighbors

Lorenz Model; 45,000 Points; T = 10; β = 0.1

FIGURE 4.15.   Percentage of bad predictions P_K as a function of local dimension d_L and 'noise level' at $N_B = 50$ for time delay vector data from the $x(t)$ dynamical variable from the Lorenz attractor, Equation (3.5). A time delay of $T = 10$ was used as determined by average mutual information. Forty five thousand data points were used, and the horizon of bad predictions was set at βR_A with $\beta = 0.1$. The noise was uniform random numbers in the interval $[-L, L]$ and the noise level is given as a fraction of the RMS size R_A of the Lorenz attractor. At about 5% global noise on the system, we lose the ability to distinguish $d_L = 3$ for this system.

4.9   A Few Remarks About Local False Nearest Neighbors

The determination of the local dimension of the dynamics by the local false nearest neighbor test gives us a critical piece of information: **it tells us how many dimensions we should use to model the dynamics for purposes of prediction or control**. It also tells us how many true Lyapunov exponents we should evaluate for the system. Since Lyapunov exponents are invariants of the attractor, they serve to characterize the system as well as give us an indication of the predictability of any model we might make.

Local false nearest neighbors are a mixture of geometry and local predictability, or more precisely local model building, and thus delve into more of the dynamics than global false nearest neighbors which rests on geometry alone. It is likely possible that more powerful versions of a local false nearest neighbors algorithm can be constructed using further aspects of the

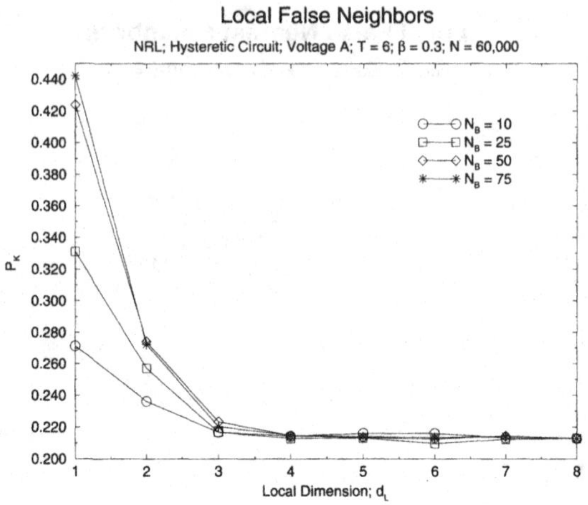

Local False Neighbors

NRL; Hysteretic Circuit; Voltage A; T = 6; β = 0.3; N = 60,000

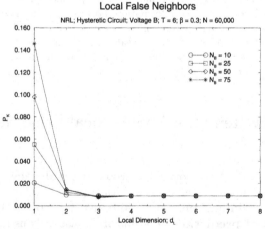

FIGURE 4.16.   Percentage of bad predictions P_K for data from voltage $V_A(t)$ of the nonlinear circuit with hysteresis from the US Naval Research Laboratory. The time delay $T = 6$ was chosen from average mutual information, and the bad prediction horizon was set at βR_A with $\beta = 0.3$. Sixty thousand data points were used. $d_L = 3$ is selected by local false nearest neighbors, while $d_E = 5$ was required by global false nearest neighbors to unfold the attractor.

Local False Neighbors

NRL; Hysteretic Circuit; Voltage B; T = 6; β = 0.3; N = 60,000

FIGURE 4.17.   Percentage of bad predictions P_K for data from voltage $V_B(t)$ of the nonlinear circuit with hysteresis from the US Naval Research Laboratory. The time delay $T = 6$ was chosen from average mutual information, and the bad prediction horizon was set at βR_A with $\beta = 0.3$. Sixty thousand data points were used. $d_L = 3$ is selected by local false nearest neighbors, and $d_E = 3$ was required by global false nearest neighbors to unfold the attractor. The consistency of $d_L = 3$ for both voltages $V_A(t)$ and $V_B(t)$ is important even though they and their time delays define different global coordinate systems and require different global embedding dimensions d_E.

local geometry and dynamics as we learn more about accurate modeling on attractors.

The local false nearest neighbors method for determining d_L is rather robust against contamination, especially when one considers that the noise levels quoted in Figure 4.15 are relative to the global size of the attractor, not relative to the size of a local neighborhood which is typically 1% or less of R_A. Local false nearest neighbors may be more robust against noise as a means of selecting d_L than forward and backward local Lyapunov exponents. The latter requires both high numerical accuracy because of ill conditionedness of matrices and very accurate evaluation of distances in phase space for evaluating local phase space Jacobians of the dynamics. Neither local nor global false nearest neighbors rests on distance evaluations for their accuracy, so they will fail less rapidly even when "noise" ruins such distance calculations.

5

Invariants of the Motion

5.1 Invariant Characteristics of the Dynamics

Classifying the dynamical systems that one observes is a critical part of the analysis of measured signals. When the source of the signals is linear, we are used to the idea of identifying the spectral peaks in the Fourier spectrum of the signal. When we see a spectral peak at some frequency, we know that if we were to stimulate the same system at a different time with forcing of a different strength, we would see the spectral peak in the same location with possibly a different integrated power under the peak. The Fourier frequency is an **invariant** of the system motion. The phase associated with that frequency depends on the time at which the measurements begin, and the power under the peak depends on the strength of the forcing. Since chaotic motion produces continuous, broadband Fourier spectra, we clearly have to replace narrowband Fourier signatures with other characteristics of the system for purposes of identification and classification. The two major features which have emerged as classifiers are **fractal dimensions** and **Lyapunov exponents**. Fractal dimensions are characteristic of the geometric figure of the attractor and relate to the way points on the attractor are distributed in d_E-dimensional space. Lyapunov exponents tell how orbits on the attractor move apart (or together) under the evolution of the dynamics. Both are invariant under the evolution operator of the system, and thus are independent of changes in the initial conditions of the orbit, and both are independent of the coordinate system in which the attractor is observed. This means we can evaluate them reliably in the reconstructed phase space made out of time delay vectors $\mathbf{y}(n)$ as described above. This is good news as it implies we can evaluate them from experimental data.

By invariants of the system dynamics we may mean something different than mathematicians speaking about dynamical systems. The Lyapunov exponents, for example, are known by the multiplicative ergodic theorem to be unchanged under smooth nonlinear changes of coordinate system, and the same is true of fractal dimensions. This is certainly not true of Fourier frequencies as nonlinear changes of coordinate will introduce new frequencies not seen in the original coordinates. The main focus when we use the word invariant in this text has to do with lack of sensitivity to initial conditions, in direct contrast with the particular orbits of the system observed or computed. This property alone is what we shall emphasize, even though some invariants are associated with operations other than changing initial conditions.

Each is connected with an ergodic theorem [ER85] which allows them to be seen as **statistical quantities characteristic of a deterministic system**. If this seems contradictory, it is only semantic. Once one has a distribution of points in R^d, as we do on strange attractors, then using the natural distribution of these points in d-dimensional space

$$\rho(\mathbf{x}) = \lim_{N\to\infty} \frac{1}{N} \sum_{k=1}^{N} \delta^d(\mathbf{x} - \mathbf{y}(k)), \tag{5.1}$$

we can define statistical quantities with this $\rho(\mathbf{x})$ acting as a density of probability or, better, frequency of occurrence. This $\rho(\mathbf{x})$ is called the **natural distribution or natural measure** since its connection to the number of points in a volume is simple. The fraction of points within a volume V of phase space is

$$\int_V d^dx\, \rho(\mathbf{x}). \tag{5.2}$$

In nonlinear systems without any noise, there are many invariant measures. All of them except this one are associated with unstable motions, so this is the only one which survives in any real experiment [ER85].

Using this density we can easily see that any function $f(\mathbf{x})$ on the phase space can be used to define an **invariant** under the evolution $\mathbf{y}(k) \to \mathbf{F}(\mathbf{y}(k)) = \mathbf{y}(k+1)$. Indeed, just integrate the function with $\rho(\mathbf{x})$ to form

$$\begin{aligned} \bar{f} &= \int d^dx\, \rho(\mathbf{x})f(\mathbf{x}), \\ &= \frac{1}{N} \sum_{k=1}^{N} f(\mathbf{y}(k)), \\ &= \frac{1}{N} \sum_{k=1}^{N} f(\mathbf{F}^{k-1}(\mathbf{y}(1))), \end{aligned} \tag{5.3}$$

where

$$\mathbf{F}^k(\mathbf{x}) = \mathbf{F}(\mathbf{F}^{k-1}(\mathbf{x})) \tag{5.4}$$

is the kth iterate of the dynamical rule $\mathbf{x} \to \mathbf{F}(\mathbf{x})$. $\mathbf{F}^0(\mathbf{x}) = \mathbf{x}$. The number
of data points N is considered simply "large", since the limit $N \to \infty$
indicated in Equation (5.1) is purely symbolic when we deal with real data.
It is easy to see that if we evaluate the average of $f(\mathbf{F}(\mathbf{x}))$, namely, the
function evaluated at the point to which $\mathbf{x}$ evolves under $\mathbf{x} \to \mathbf{F}(\mathbf{x})$, we
find

$$\int d^d x\, \rho(\mathbf{x}) f(\mathbf{F}(\mathbf{x})) = \bar{f} + \frac{1}{N}[f(\mathbf{y}(N+1)) - f(\mathbf{y}(1))], \tag{5.5}$$

which in the limit of large N is just $\bar{f}$. This invariance is so far just an interesting observation, which becomes useful only when we select the function
$f(\mathbf{x})$ well.

5.2 Fractal Dimensions

Perhaps the most interesting function $f(\mathbf{x})$ which is widely considered is
one which tells us the way in which the number of points within a sphere of
radius r scales as the radius shrinks to zero. The geometric relevance of this
is that the volume occupied by a sphere of radius r in dimension d behaves
as r^d, so we might expect to achieve a sense of dimension by seeing how
the density of points on an attractor scales when we examine it at small
distances in phase space.

To motivate this a bit more, imagine a set of points in some Euclidian
space, namely, the embedding space constructed from the data $\mathbf{y}(n); n = 1, 2, \ldots, N$. Go to some point $\mathbf{x}$ on or near the attractor and ask how the
number of points on the orbit within a distance r of $\mathbf{x}$ changes as we make
r small. Well, not too small, since we have only a finite number of data,
and soon the radius will be so small that no points fall within the sphere.
So, small enough that there are a lot of points within the sphere and not
so large that all the data is within the sphere. The latter would happen if
$r \approx R_A$. In this intermediate regime of r we expect the number of points
$n(\mathbf{x}, r)$ within r at $\mathbf{x}$ to scale as

$$n(\mathbf{x}, r) \approx r^{d(\mathbf{x})}, \tag{5.6}$$

for small r. "Small" should really be stated in dimensionless terms, and the
only scale with which we can naturally measure r is the attractor size R_A.
We always mean

$$0 \ll \frac{r}{R_A} \ll 1, \tag{5.7}$$

when we say "small r".

If the attractor were a regular geometric figure of dimension D, then
in each such ball of radius r around $\mathbf{x}$ we would find approximately r^D

points—times some overall numerical geometric factor of no special importance here. This would lead us to set $d(\mathbf{x}) = D$ for all $\mathbf{x}$. On a chaotic or
strange attractor, which is a geometric figure not as regular as a sphere
or a torus, we do not expect to get the same value for D everywhere, so
having $d(\mathbf{x})$ vary with $\mathbf{x}$ is natural [Ott93, PV87]. We might call $d(\mathbf{x})$ a
local dimension, but since it refers to some specific point on the attractor,
we have no particular reason to think it would be the same for all $\mathbf{x}$, and
thus under the action of the dynamics $\mathbf{x} \to \mathbf{F}(\mathbf{x})$, it would change. It is
sensitive to initial conditions. To find something invariant under $\mathbf{F}(\mathbf{x})$ we
need to average a given local function $f(\mathbf{x})$ with the natural invariant measure $\rho(\mathbf{x})$. For the purpose of identifying dimensions in this fashion, we
turn our attention to the number of observed points $\mathbf{y}(k)$ within a sphere
around some phase space location $\mathbf{x}$. This is found by

$$n(\mathbf{x},r) = \frac{1}{N} \sum_{k=1}^{N} \theta(r - |\mathbf{y}(k) - \mathbf{x}|), \qquad (5.8)$$

where $\theta(u)$ is the Heaviside function

$$\begin{aligned} \theta(u) &= 1 \ \text{if} \ u > 0, \\ &= 0 \ \text{if} \ u < 0. \end{aligned} \qquad (5.9)$$

This counts all the points on the orbit $\mathbf{y}(k)$ within a radius of r from the
point $\mathbf{x}$ and normalizes that number by the total number of data points.
Now we recognize that the density of $\rho(\mathbf{x})$ points on an attractor need not
be uniform on the figure of the attractor, so it may be quite revealing to
look at the moments of the function $n(\mathbf{x}, r)$. The density $\rho(\mathbf{x})$ is inhomo-
geneous on a strange attractor in general, so the moments of any function
$f(\mathbf{x})$ evaluated with this natural density will reveal different aspects of the
distribution of points.

We choose for our function $f(\mathbf{x}) = n(\mathbf{x},r)^{(q-1)}$ and define the function
$C(q,r)$ of two variables q and r by the mean of $f(\mathbf{x})$ over the attractor
weighted with the natural density $\rho(\mathbf{x})$:

$$\begin{aligned} C(q,r) &= \int d^d x \, \rho(\mathbf{x}) n(\mathbf{x},r)^{(q-1)}, \\ &= \frac{1}{M} \sum_{k=1}^{M} \left[\frac{1}{K} \sum_{n=1}^{K} \theta(r - |\mathbf{y}(n) - \mathbf{y}(k)|) \right]^{(q-1)}. \end{aligned} \qquad (5.10)$$

This is often called the "correlation function" on the attractor. While such
quantities were known in statistics from the work of Renyi [Ren70], the
idea of examining this kind of quantity to characterize strange attractors
is due to Grassberger and Procaccia [GP83] who originally discussed the
case $q = 2$. As usual, we write M and K imagining each to be large but
not imagining they are infinite. This means we must always remember not

to literally take $r/R_A \to 0$, for there will be no points to be counted in evaluating $C(q,r)$.

This whole function of two variables is an invariant on the attractor, but it has become conventional to look only at the variation of this quantity when r is small. In that limit it is *assumed* that

$$C(q,r) \approx r^{(q-1)D_q}, \tag{5.11}$$

defining the fractal dimension D_q, when it exists. Clearly there is a geometric appeal to seeking a single number for each moment of the density of points in the sphere of radius r for small r, but we really need not think of this as fundamental. We shall see in just a moment that D_q is defined by a limit, and because of this one can show it is invariant under changes in coordinate system, so it takes on a particularly attractive geometric meaning. However, if one is working in a given coordinate system, defined by the time delay vectors $\mathbf{y}(n)$ in a fixed embedding dimension d_E, then the whole curve $C(q,r)$ becomes of interest. Since this is the situation one encounters when analyzing data from a given observed source, it seems wise not to throw away the full information in $C(q,r)$ just to focus on the slope of a hoped for linear segment of it when $\log[C(q,r)]$ is plotted against $\log[r]$.

The dimension D_q is defined by a limit

$$D_q = \lim_{r \text{ small}} \frac{\log[C(q,r)]}{(q-1)\log[r]}, \tag{5.12}$$

and from this we can see that the overall normalization of $C(q,r)$ doesn't matter for the evaluation of D_q. Also we learn that $D_{q-1} \geq D_q$.

In practice we need to compute $C(q,r)$ for a range of small r over which we can argue that the function $\log[C(q,r)]$ is linear in $\log[r]$ and then pick off the slope over that range. This is not as easy as it sounds, and numerous papers have been written on how one does this with a finite amount of noisy data and what to do about $r \neq 0$, etc. Essentially all the work has concentrated on the quantity D_2 because its evaluation is numerically rather simple and reliable. The papers by Theiler [The90], Smith [Smi88], Ruelle [Rue90], and Essex and Nerenberg [EN91] are critical of how one makes these calculations, what one believes about these calculations, how much data is required for these calculations, and other issues as well. It is out of the question to discuss all that has been said and written on this subject, but a flavor of the discussion can be gotten from those papers. One of the interesting points is the rule of thumb that, if one is to evaluate D_2 with some confidence, then a decade of dynamic range in $\log[r]$ is required. This suggests that at least $10^{D_2/2}$ data points are needed to believe a report of a fractal dimension D_2. While only a rule of thumb, it is a useful one. The rule of thumb can be used to sound a warning about results found in examining data. If a computation yields a value for D_2 which is near or less than $2\log_{10} N$ for N data points, natural skepticism about the result is called for.

While as a matter of course I will report some values of D_2 evaluated by the method suggested above, it is well worth examining what one actually learns by evaluating D_2. Since we would have established by now using false nearest neighbors that we have a low dimensional system to work with, and we would know from the same method what dimension we require to unfold the attractor, we would then be seeking a single number to characterize the attractor. If D_2 is not integer, this is a very interesting statement about the dynamics at the source of our observed signal, but certainly not a complete characterization of the source. Indeed, at this time no one knows what constitutes a complete set of invariants to characterize an attractor. Further we would have long since established that the source of our observations is low dimensional so that a small D_2 is not news.

As long as we do not think of the evaluation of D_2 as a way to determine whether the signal comes from a low dimensional source or whether the signal comes from a deterministic source but simply as a characteristic number for that source, we will be on stable ground. It is important to ask just what we will have learned about the source of the chaotic signal from this single number, especially with the remarkable uncertainty that enters its determination from experimental data.

Now this has been a rather long introduction. Absent the fact that so much importance has been placed on the dimensions D_q, especially D_2, it would not have been warranted. Given the huge effort placed on the evaluation of this single quantity and the importance placed on its interpretation, it seems worth making this alert ahead of any numerical display of its values. Personally I am inclined to place more interest in the whole function $C(q,r)$ and its values for a wide range of r. When the assumption that $C(q,r)$ behaves as a power of r fails, the function itself may still be quite interesting.

5.2.1 D_0: Box Counting

One of these dimensions, namely, D_0, has a nice clean motivation. It is called the "box counting" dimension because of its definition. To evaluate D_0 we ask how many spheres of radius r, namely, how many boxes, do we need to cover all the points in the data set. If we evaluate this number, $N(r)$ as a function of r as it becomes small, then the ratio

$$D_0 = \lim_{r \to 0} \frac{\log N(r)}{\log \frac{1}{r}}, \qquad (5.13)$$

defines D_0 and quantifies the idea that for small enough r,

$$N(r) \approx r^{-D_0}. \qquad (5.14)$$

Simple examples support this as a definition of dimension. Suppose our set is a piece of line segment, and we cover it with line segments of length r.

One piece might be of length $c_1 r$, and another piece might be of length $c_2 r$, and yet another of length $c_3 r$, etc., with the c_i constants of order unity. The number of such segments would be of order C/r, with C another finite constant. This gives $D_0 = 1$, independently of the value of C, as it must. Similarly covering an area with small circles of radius r requires C/r^2 of those circles when r^2 is much less than the area itself.

The definition of box counting dimension is not restricted to familiar integer examples. We can establish a fractional dimension for a set recognizing that we need be careful as the dimension of any finite set of points is always zero, but we never have an infinite number of points in practice, only in our mathematical examples. In particular this box counting idea immediately allows us to find the dimension of the time honored "middle third" Cantor set [Can83] and show it is not an integer. To construct this infinite set of points, start with a unit line segment and remove the part between $1/3$ and $2/3$. Then from each of these remaining segments, remove its middle third, leaving after this second step four line segments $[0, 1/9], [2/9, 1/3], [2/3, 7/9]$ and $[8/9, 1]$. After m such operations we have 2^m pieces of line segment, each of length 3^{-m}. This defines the number of boxes, $N(r) = 2^m$, of radius $r = 1/3^m$ that we need, and immediately tells us that

$$
\begin{aligned}
D_0 &= \lim_{m \to \infty} \frac{\log 2^m}{\log 3^m} \\
&= \frac{\log 2}{\log 3} \\
&\approx 0.6309297 \\
0 < \ & D_0 \ < 1.
\end{aligned}
\tag{5.15}
$$

The box counting dimension of the Cantor set is fractional, indeed irrational, and less than one. The infinite number of points generated as the number of middle third removing operations goes to infinity constitute more than a finite set of points which would have dimension zero. They also constitute less than all the points in the original line segment which we argued has dimension one. We also note that the quantity d_A which we used as "the" dimension of the attractor in discussing the embedding process is identified as $d_A = D_0$ [CSY91]. This is natural since it is the absence of overlap of two sets of dimension d_A which leads to the sufficient requirement that $d_E > 2d_A$ for unfolding an attractor of dimension d_A. Overlap is a geometric concept clearly connected with covering sets with boxes of size r.

5.2.2 Lorenz Model

We choose to display the function $C(2, r)$ for the Lorenz model using quite a bit of well sampled, clean data. Each computation used 50,000 points

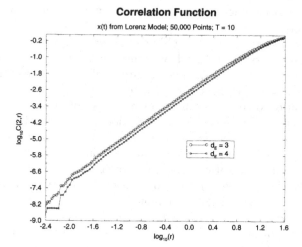

FIGURE 5.1.   The correlation function $C(2, r)$ as a function of r for data from the $x(t)$ component of the Lorenz attractor, Equation (3.5). Fifty thousand data points were used and $T = 10$ selected by average mutual information was utilized in making time delay vectors for spaces of dimension $d_E = 3$ and 4. We already know from false nearest neighbors that this attractor is unfolded in $d_E = 3$, so $d_E = 4$ is a consistency check. The slope of this graph yields the correlation dimension D_2.

from the Lorenz system with $\tau_s = 0.01$. In these dimensionless units an approximate time to go around the attractor is 0.5. 50,000 points means circulating about the attractor nearly 1000 times. This is not what one would usually find in observed data, so this is a departure from the tone of this book, but it may be useful to see what a clean example would produce. In Figures 5.1, 5.2, and 5.3 we display the curves $\log_{10}[C(2, r)]$ versus $\log_{10}[r]$ evaluated from each of the three variables $x(n), y(n)$, and $z(n)$. The curves are shown using vectors in both $d_E = 3$ and $d_E = 4$. The normalization of the function $C(q, r)$ depends on the embedding dimension, but, accounting for that, one sees that the various curves are the same within numerical precision. This is as it should be for invariants and emphasizes the utility both of the whole function $C(q, r)$ and the strict invariance of its logarithm with respect to $\log[r]$. The Figures 5.4, 5.5, and 5.6 are a plot of

$$\frac{d \log[C(2, r)]}{d \log[r]}, \tag{5.16}$$

versus $\log[r]$ for each of the data sets $x(n), y(n)$, and $z(n)$ from the Lorenz system. Derivatives are defined as the local average over three neighboring points. While one can see a "middle" region in $\log[r]$ where these derivatives are all just above two, the difficulties of establishing a clean, unsullied value for D_2 should be clear. With real data it doesn't get better.

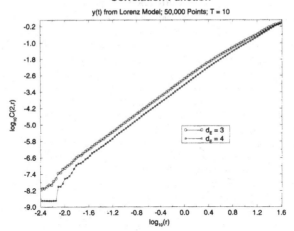

Correlation Function

y(t) from Lorenz Model; 50,000 Points; T = 10

FIGURE 5.2.   The correlation function $C(2, r)$ as a function of r for data from the $y(t)$ component of the Lorenz attractor, Equation (3.5). Fifty thousand data points were used and $T = 10$ selected by average mutual information was utilized in making time delay vectors for spaces of dimension $d_E = 3$ and 4. We already know from false nearest neighbors that this attractor is unfolded in $d_E = 3$, so $d_E = 4$ is a consistency check. The slope of this graph yields the correlation dimension D_2.

5.3   Global Lyapunov Exponents

The stability of an observed orbit $\mathbf{y}(k)$ of the dynamical system $\mathbf{y}(k) \rightarrow \mathbf{F}(\mathbf{y}(k)) = \mathbf{y}(k + 1)$ to small perturbations $\Delta(k)$ is revealed by looking at the linearized dynamics

$$\begin{aligned} \mathbf{y}(k+1) + \Delta(k+1) &= \mathbf{F}(\mathbf{y}(k) + \Delta(k)) \\ &\approx \mathbf{DF}(\mathbf{y}(k)) \cdot \Delta(k) + \mathbf{F}(\mathbf{y}(k)), \quad (5.17) \end{aligned}$$

or

$$\Delta(k+1) = \mathbf{DF}(\mathbf{y}(k)) \cdot \Delta(k), \quad (5.18)$$

as long as $\Delta(\bullet)$ remains small. In this the Jacobian matrix

$$DF(\mathbf{x})_{ab} = \frac{\partial F_a(\mathbf{x})}{\partial x_b} \quad (5.19)$$

enters. The stability of the orbit is determined by the fate of $\Delta(k)$ as the number of evolution steps from the starting time k grows large. Suppose we move ahead from time k to time $k + L$ using the linearized evolution for $\Delta(k)$. Then we find

$$\begin{aligned} \Delta(k+L) &= \mathbf{DF}(\mathbf{y}(k+L-1)) \cdot \mathbf{DF}(\mathbf{y}(K+L-2)) \cdots \mathbf{DF}(\mathbf{y}(k)) \cdot \Delta(k) \\ &\equiv \mathbf{DF}^L(\mathbf{y}(k)) \cdot \Delta(k), \quad (5.20) \end{aligned}$$

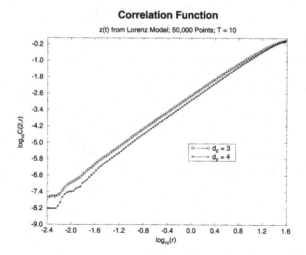

FIGURE 5.3.   The correlation function $C(2, r)$ as a function of r for data from the $z(t)$ component of the Lorenz attractor, Equation (3.5). Fifty thousand data points were used and $T = 10$ selected by average mutual information was utilized in making time delay vectors for spaces of dimension $d_E = 3$ and 4. We already know from false nearest neighbors that this attractor is unfolded in $d_E = 3$, so $d_E = 4$ is a consistency check. The slope of this graph yields the correlation dimension D_2.

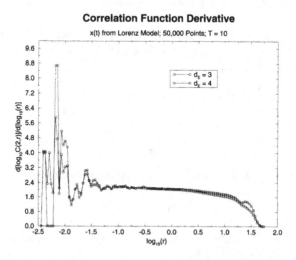

FIGURE 5.4.   The derivative of the correlation function $\log[C(2, r)]$ created from $x(t)$ taken from the Lorenz model, Equation (3.5), with respect to $\log(r)$ evaluate for $d_E = 3$ and $d_E = 4$. We see consistency in a broad range of $\log(r)$ with a slope slightly larger than two.

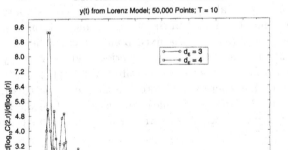

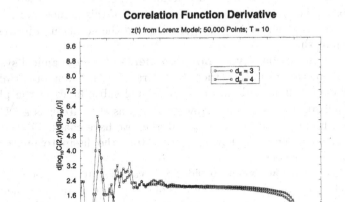

FIGURE 5.5.   The derivative of the correlation function $\log[C(2,r)]$ created from $y(t)$ taken from the Lorenz model, Equation (3.5) with respect to $\log(r)$ evaluate for $d_E = 3$ and $d_E = 4$. We see consistency in a broad range of $\log(r)$ with a slope slightly larger than two.

FIGURE 5.6.   The derivative of the correlation function $\log[C(2,r)]$ created from $z(t)$ taken from the Lorenz model, Equation (3.5), with respect to $\log(r)$ evaluate for $d_E = 3$ and $d_E = 4$. We see consistency in a broad range of $\log(r)$ with a slope slightly larger than two.

defining $\mathbf{DF}^L(\mathbf{x})$ as the composition of L Jacobians. In an intuitive sense we see that if the eigenvalues of $\mathbf{DF}^L(\mathbf{y}(k))$ behave as $\exp[L\lambda]$ with $\lambda > 0$, then the orbit $\mathbf{y}(k)$ along which the Jacobians are evaluated is unstable. This is the generalization from the standard linear stability [Dra92, Cha61] under small perturbations of a fixed point which is the case when $\mathbf{y}(k)$ is independent of time. When $\mathbf{y}(k)$ is periodic $\mathbf{y}(k + P) = \mathbf{y}(k)$, then the usual Floquet theory governs the stability. When $\mathbf{y}(k)$ is a time dependent chaotic orbit, then there is a theorem associated with this problem which is due to the Russian mathematician Oseledec [Ose68]. It is called the **multiplicative ergodic theorem** and states that if we look at the square of the length of the vector $\Delta(k + L)$,

$$|\Delta(k + L)|^2 = \Delta^T(k) \cdot \left[\mathbf{DF}^L(\mathbf{x})\right]^T \cdot \mathbf{DF}^L(\mathbf{x}) \cdot \Delta(k), \qquad (5.21)$$

then the essential quantity determining this is

$$\left[\mathbf{DF}^L(\mathbf{x})\right]^T \cdot \mathbf{DF}^L(\mathbf{x}), \qquad (5.22)$$

where the superscript T means transpose.

The multiplicative ergodic theorem states that if we form the Oseledec matrix

$$\mathbf{OSL}(\mathbf{x}, L) = (\left[\mathbf{DF}^L(\mathbf{x})\right]^T \cdot \mathbf{DF}^L(\mathbf{x}))^{\frac{1}{2L}}, \qquad (5.23)$$

then the limit of this as $L \to \infty$ exists and is independent of $\mathbf{x}$ for all $\mathbf{x}$ (well, for almost all $\mathbf{x}$) in the basin of attraction of the attractor to which the orbit belongs. The logarithm of the eigenvalues of this orthogonal matrix when $L \to \infty$ are denoted $\lambda_1 \geq \lambda_2 \geq, \ldots, \lambda_d$ and the notation indicates we order them as shown.

The λ_a are the **global Lyapunov exponents** of the dynamical system $\mathbf{x} \to \mathbf{F}(\mathbf{x})$. One can also define them by the rate of stretching or shrinkage of line segments, area, and various dimensional subvolumes in the phase space [BGGS80]. Line segments grow or shrink as $e^{t\lambda_1}$, areas as $e^{t(\lambda_1 + \lambda_2)}$, as so forth. If one or more of the $\lambda_a > 0$, then we have chaos. The sum of the Lyapunov exponents $\lambda_1 + \lambda_2 +, \cdots, \lambda_d < 0$ by the dissipative nature of the systems we consider.

As we discuss below, when considering local Lyapunov exponents, the λ_a are unchanged under smooth changes of coordinate system, so one may evaluate them in any coordinate system one chooses. In particular, the value for the λ_a as evaluated in the original coordinates for the system **or** in the reconstruction coordinates provided by time delays of any measured quantity, are the same. The λ_a are also unchanged under the dynamics $\mathbf{x} \to \mathbf{F}(\mathbf{x})$ when the vector field $\mathbf{F}(\bullet)$ is smooth, as this is essentially a change of coordinates.

The determination of the eigenvalues λ_a of the Oseledec matrix is not numerically trivial even though the dimension of the matrix may be small. The point is that $\left[\mathbf{DF}^L(\mathbf{x})\right]^T \cdot \mathbf{DF}^L(\mathbf{x})$ is quite ill-conditioned [GV89] as

L becomes large. The condition number is approximately $e^{[L(\lambda_1-\lambda_d)]}$. The evaluation of the eigenvalues rests on the idea of a recursive QR decomposition which was described by Eckmann, *et al* [EKRC86] and works for large L.

This recursive QR decomposition treats the problem associated with the ill-conditioned nature of the matrices entering $\mathbf{OSL}(\mathbf{x}, L)$ for large L by effectively partially diagonalizing the "large part" of the matrix step by step. The Oseledec matrix to the power $2L$ is a product of factors which we can represent as $\mathbf{A}(2L) \cdot \mathbf{A}(2L-1), \cdots, \mathbf{A}(1)$. Any component matrix $\mathbf{A}(j)$ in this product can always be written as a product of an orthogonal matrix $\mathbf{Q}(j)$ and an upper right triangular matrix $\mathbf{R}(j)$. This is analogous to a polar decomposition of a complex number which writes the complex number as a radius times a phase factor. The orthogonal matrix $\mathbf{Q}$ is like the phase factor and the $\mathbf{R}$ like the magnitude of the complex number. $\mathbf{R}$ is the part of the matrix which becomes large and leads to the overall ill-conditioned behavior of $\mathbf{OSL}$.

The idea is to write each $\mathbf{A}(j)$ as

$$\mathbf{A}(j) \cdot \mathbf{Q}(j-1) = \mathbf{Q}(j) \cdot \mathbf{R}(j), \qquad (5.24)$$

where $\mathbf{Q}(0) = \mathcal{I}$, the identity matrix. This would give for the first part of the product

$$\begin{aligned}
\mathbf{A}(1) &= \mathbf{Q}(1) \cdot \mathbf{R}(1), \\
\mathbf{A}(2) \cdot \mathbf{Q}(1) &= \mathbf{Q}(2) \cdot \mathbf{R}(2), \\
\mathbf{A}(2) &= \mathbf{Q}(2) \cdot \mathbf{R}(1) \cdot \mathbf{Q}(1)^T, \\
\mathbf{A}(2) \cdot \mathbf{A}(1) &= \mathbf{Q}(2) \cdot \mathbf{R}(2) \cdot \mathbf{R}(1), \qquad (5.25)
\end{aligned}$$

and the next step would be

$$\mathbf{A}(3) \cdot \mathbf{A}(2) \cdot \mathbf{A}(1) = \mathbf{Q}(2) \cdot \mathbf{R}(3) \cdot \mathbf{R}(2) \cdot \mathbf{R}(1), \qquad (5.26)$$

and for the full product

$$\mathbf{A}(2L) \cdot \mathbf{A}(2L-1), \cdots, \mathbf{A}(1) = \mathbf{Q}(2L) \cdot \mathbf{R}(2L) \cdot \mathbf{R}(2L-1) \cdots \mathbf{R}(1). \quad (5.27)$$

This is easily diagonalized, as the product of upper right triangular matrices is an upper right triangular matrix, and the eigenvalues of such a matrix are the numbers along the diagonal. The Lyapunov exponent is read off the product of the upper triangular matrices as

$$\lambda_a = \lim_{L\to\infty} \frac{1}{2L} \sum_{k=1}^{2L} \log[R_{aa}(k)]. \qquad (5.28)$$

5.4   Lyapunov Dimension

The λ_a give us a sense of dimension [Ott93, KY79]. When we have a chaotic system, there is at least one positive Lyapunov exponent. This is the signal of the intrinsic instability we call chaos. The largest of the exponents λ_1 determines how line segments grow under the dynamics. Areas grow according to $e^{[L(\lambda_1+\lambda_2)]}$. Three-dimensional volumes grow according to $e^{[L(\lambda_1+\lambda_2+\lambda_3)]}$, etc. A volume in d-dimensional space behaves as $e^{[L(\lambda_1+\lambda_2+...+\lambda_d)]}$, so the sum of all exponents must be negative. Some combination of exponents can be associated with a volume in phase space which neither grows nor shrinks.

Kaplan and Yorke [KY79] suggested that this be used to define a Lyapunov dimension

$$D_L = K + \frac{\sum_{a=1}^{K} \lambda_a}{|\lambda_{K+1}|}, \tag{5.29}$$

where $\sum_{a=1}^{K} \lambda_a > 0$ and $\sum_{a=1}^{K+1} \lambda_a < 0$. This dimension has been associated with the information dimension D_1, as defined above, but no general connection seems to have been made to the satisfaction of mathematicians. This is not a big problem in any practical sense since D_L is generally about the same size as the D_q we discussed earlier.

To motivate the idea of the Lyapunov dimension, we consider the evolution in time of an initial sphere of radius r in d dimensions. This becomes distorted to an ellipse with axes of length approximately $re^{\lambda_a t}$. Choose an integer $K < d$ for which $\lambda_{K+1} < 0$ and ask how many boxes of size $re^{\lambda_{K+1}t}$ are required to capture the K-dimensional volume

$$e^{\lambda_1 t} e^{\lambda_2 t}, \ldots, e^{\lambda_K t} r^K. \tag{5.30}$$

Since we have chosen an exponent $\lambda_{K+1} < 0$, we need not consider the contribution of directions which shrink faster than $re^{\lambda_{K+1}t}$. The number of required boxes of size $re^{\lambda_{K+1}t}$ is

$$N(t) = \frac{e^{t(\lambda_1+\lambda_2+...\lambda_K)}}{e^{Kt\lambda_{K+1}}}. \tag{5.31}$$

The Lyapunov dimension is defined by the box counting rule to be

$$\begin{aligned} D_L &= \lim_{t\to\infty} \frac{\log N(t)}{\log(1/e^{t\lambda_{K+1}})} \\ &= K + \frac{\sum_{a=1}^{K} \lambda_a}{|\lambda_{K+1}|}. \end{aligned} \tag{5.32}$$

This quantity is minimized when K is chosen such that $\sum_{a=1}^{K} \lambda_a > 0$ and $\sum_{a=1}^{K+1} \lambda_a < 0$. The minimum gives us the smallest dimensioned subspace in which the set of points on the orbit can be captured. It is natural to identify this as a fractal dimension for the attractor. When one is able to determine the spectrum of global Lyapunov exponents λ_a, the dimension D_L serves numerically quite well as an estimates of the D_q for q a small integer. The downside of evaluating D_L rather than D_q is that one requires some negative Lyapunov exponents, and with noisy data these may be difficult to determine reliably.

5.5   Global Lyapunov Exponents from Data

To find the Lyapunov exponents from observed scalar data we need some method for accurately determining the Jacobian matrix $\mathbf{DF}(\mathbf{y}(k))$ at locations on the attractor which are visited by the orbit $\mathbf{y}(k)$. For this we require some way to acquire a sense of the variation of the dynamics $\mathbf{y}(k+1) = \mathbf{F}(\mathbf{y}(k))$ in the neighborhood of the observed orbit. The main idea on how we can do this [EKRC86, SS85] is to recognize that attractors are compact objects in their phase space and that any orbit will come back into the neighborhood of any given point on the attractor given a long enough orbit. Thus from one orbit we can acquire information about the phase space behavior of quantities such as $\mathbf{F}(\bullet)$ by looking at the phase space neighbors of $\mathbf{y}(k)$.

Let us look at $\mathbf{y}(k)$ and find its N_B nearest neighbors: $\mathbf{y}^{(r)}(k); r = 1, 2, \ldots, N_B$. Each of these neighbors evolves into a known point $\mathbf{y}^{(r)}(k) \rightarrow \mathbf{y}(r; k+1)$ which is in the neighborhood of $\mathbf{y}(k+1)$. The notation is chosen to emphasize that $\mathbf{y}(r; k+1)$ may not be the rth nearest neighbor $\mathbf{y}^{(r)}(k+1)$ of $\mathbf{y}(k+1)$. If we make a **local** map at 'time' k

$$\mathbf{x} \rightarrow \sum_{m=1}^{M} \mathbf{c}(m, k)\phi_m(\mathbf{x}), \qquad (5.33)$$

from neighborhood to neighborhood:

$$\mathbf{y}(r; k+1) = \sum_{m=1}^{M} \mathbf{c}(m, k)\phi_m(\mathbf{y}^{(r)}(k)), \qquad (5.34)$$

where the functions $\phi_m(\mathbf{x})$ are some basis set we choose *a priori*, and the $\mathbf{c}(m, k)$ are local coefficients we will determine in a moment, then the components of the desired Jacobian matrix are

$$DF_{ab}(\mathbf{y}(k)) = \sum_{m=1}^{M} c_a(m, k)\frac{\partial \phi_m(\mathbf{x})}{\partial x_b}, \qquad (5.35)$$

evaluated at $\mathbf{x} = \mathbf{y}(k)$. If the local functions are polynomials [BBA91, Bri90], then the linear term in the mapping determines $\mathbf{DF}(\mathbf{y}(k))$.

Retaining only the linear term in the polynomial expression for the local map puts a severe numerical burden on the local maps. It requires them both to follow the twists and turns of the orbits along the attractor and to produce accurate numerical values for the local Jacobians. By using higher order polynomials the other terms have the role of making the local mapping accurate while the linear term is still used to determine the Jacobian to use for determining Lyapunov exponents. If there are inaccuracies in the evaluation of $\mathbf{DF}(\mathbf{y}(k))$, they tend to be exponentially magnified by the ill-conditioned nature of the numerical problem here. Ill-conditionedness means that the eigenvalues of the matrix have a very large ratio with respect to each other, and this requires numerical accuracy of exponential order to produce good results for all eigenvalues. Here the eigenvalues of the elements in the Oseledec matrix are of order $e^{\lambda_a L}$. The ratio of the first two eigenvalues is

$$\epsilon^{(\lambda_1 - \lambda_2)L}. \tag{5.36}$$

For large L, since $\lambda_1 > \lambda_2$, this goes to infinity very rapidly.

Polynomials are only one choice [Par92, ABST93, Pow81] as basis function in which to express the local map from neighborhood to neighborhood. The coefficients $\mathbf{c}(m, k)$ in Equation (5.34) are determined by requiring the residuals

$$\sum_{r=1}^{N_B} |\mathbf{y}(r, k+1) - \sum_{m=1}^{M} \mathbf{c}(m, k)\phi_m(\mathbf{y}^{(r)}(k))|^2, \tag{5.37}$$

to be minimized. The determination of the coefficients is made at each data point from information in each neighborhood. The coefficients are used to evaluate the local Jacobian matrix which is then put into the Oseledec matrix to be diagonalized by the recursive QR method. Needless to say when one has a large data set and is working in dimensions as large as five or more, this task can be quite demanding on computer resources.

5.6   Local Lyapunov Exponents

The global Lyapunov exponents we discussed above tell us how a perturbation to an orbit $\mathbf{y}(k)$ will behave over a long time. This is stability information which has been averaged over the whole attractor. This may not be particularly relevant information since what happens as time gets very large may be of no importance to what happens in the next few moments. Predicting weather five days from now is certainly more interesting than 10^5 days (274 years) from now.

The eigenvalues of the Oseledec matrix $\mathbf{OSL}(\mathbf{x}, L)$

$$\exp[2L\lambda_a(\mathbf{x}, L)], \tag{5.38}$$

tell us how rapidly perturbations to the orbit at point $\mathbf{x}$ in phase space grow or shrink in L time steps away from the time of the perturbation. These $\lambda_a(\mathbf{x}, L)$ are called **local Lyapunov exponents** [ABK91, ABK92] or finite time Lyapunov exponents [GBP88]. They certainly become the global exponents for large L

$$\lambda_a(\mathbf{x}, L) \to \lambda_a, \tag{5.39}$$

as $L \to \infty$. The variations around the limit are also of importance. The $\lambda_a(\mathbf{x}, L)$ vary significantly with the location $\mathbf{x}$ on the attractor, especially for small L which is the main interest. The moments of $\lambda_a(\mathbf{x}, L)$, with moments defined by integrals with the natural density $\rho(\mathbf{x})$, are invariants of the dynamics. The average local Lyapunov exponent

$$\bar{\lambda}_a(L) = \int \rho(\mathbf{x})\lambda_a(\mathbf{x}, L) \tag{5.40}$$

satisfies

$$\bar{\lambda}(L)_a \approx \lambda_a + \frac{K_a}{L^{\nu_a}} + \frac{K'_a}{L}, \tag{5.41}$$

where K_a, K'_a and $\nu_a < 1$ are constants. The last term comes from the geometric dependence of the $\lambda_a(\mathbf{x}, L)$ on the coordinate system in which it is evaluated as we will consider in Section 5.6.2.

Moments of the local exponent around the mean $\bar{\lambda}_a(L)$ all vanish as $L \to \infty$. In particular

$$\int d^d x \rho(\mathbf{x})[\lambda_a(\mathbf{x}, L) - \bar{\lambda}_a(L)]^p \approx \frac{K''_a}{L^{\xi_a(p)}}, \tag{5.42}$$

where $\xi_a(p) \approx p\nu_a$.

There is quite a bit of information in these local exponents and in their averages around the attractor $\bar{\lambda}_a(L)$.

i First, they tell us on the average around the attractor how well we can predict the evolution of the system L steps ahead of wherever we are. If the second and higher moments are substantial, then this predictability will vary substantially as we move about the attractor.

ii Second, if we are observing data from a flow, that is, a set of differential equations, then one of the λ_a **must** be zero [ER85]. The reason is this: if we choose to make a perturbation to an orbit exactly along the direction the orbit is going, then that perturbation will be another orbit point and will move precisely along the same orbit. Divergence of the new orbit from the old will be absent; this gives $\lambda_a = 0$ for that particular direction. In the case of a mapping underlying our data, there is no flow direction. If we observe an average local Lyapunov exponent $\bar{\lambda}_a(L)$ going to zero, we can be confident that we have a flow.

iii Third, the values of the $\bar{\lambda}(L)$ are each dynamical invariants characterizing the source of the measurements.

5.6.1   Recursive QR Decomposition for Short Times

When we wish to find the local Lyapunov exponents $\lambda_a(\mathbf{x}, L)$ for small L it is necessary to modify the recursive QR decomposition of Eckmann, et al. [EKRC86] which was developed for the case of $L \to \infty$.

The quantities one wishes to compute are the eigenvalues of

$$[\mathbf{DF}^L]^T \cdot \mathbf{DF}^L = \mathbf{DF}(1)^T \cdot \mathbf{DF}(2)^T, \ldots, \tag{5.43}$$
$$\mathbf{DF}(L)^T \cdot \mathbf{DF}(L) \cdot \mathbf{DF}(L-1), \ldots, \mathbf{DF}(1),$$

which is a product of $2L$ matrices.

When we make the recursive QR decomposition of $[\mathbf{DF}^L(\mathbf{x})]^T \cdot \mathbf{DF}^L(\mathbf{x})^T$ as

$$\mathbf{Q}_1(2L) \cdot \mathbf{R}_1(2L) \cdot \mathbf{R}_1(2L-1), \ldots, \mathbf{R}_1(1) = \mathbf{M}_1, \tag{5.44}$$

if $\mathbf{Q}_1(2L)$ were the identity matrix, the eigenvalues of $\mathbf{OSL}(\mathbf{x}, L)$ would all lie in the 'upper triangular' part of the 'QR' decomposition. In general, $\mathbf{Q}_1(2L)$ is not the identity for finite L. So we shuffle this Q factor over to the right of all the R matrices, and repeat our QR decomposition.

For this we define a matrix $\mathbf{M}_2$ which has the same $\mathbf{R}$ factors as $\mathbf{M}_1$ but has the matrix $\mathbf{Q}_1(2L)$ on the right:

$$\mathbf{M}_2 = \mathbf{R}_1(2L) \cdot \mathbf{R}_1(2L-1), \ldots, \mathbf{R}_1(1) \cdot \mathbf{Q}_1(2L), \tag{5.45}$$
$$\mathbf{M}_2 = \mathbf{Q}_1^T(2L) \cdot \mathbf{M}_1 \cdot \mathbf{Q}_1(2L), \tag{5.46}$$

and then we perform the recursive QR decomposition once again on $\mathbf{M}_2$:

$$\mathbf{M}_2 = \mathbf{Q}_2(2L) \cdot \mathbf{R}_2(2L) \cdot \mathbf{R}_2(2L-1), \ldots, \mathbf{R}_2(1).$$

Since $\mathbf{M}_2 = \mathbf{Q}_1^T(2L) \cdot \mathbf{M}_1 \cdot \mathbf{Q}_1(2L)$, $\mathbf{M}_1$ and $\mathbf{M}_2$ have the same eigenvalues. Continue this sequence of operations creating $\mathbf{M}_3, \mathbf{M}_4, \ldots, \mathbf{M}_K$:

$$\mathbf{M}_K = \mathbf{Q}_K(2L) \cdot \mathbf{R}_K(2L) \cdot \mathbf{R}_K(2L-1), \ldots, \mathbf{R}_K(1). \tag{5.47}$$

A theorem of numerical analysis states that as K increases, $\mathbf{Q}_K(2L)$ converges to the identity matrix [SB80].

When $\mathbf{Q}_K(2L)$ is the identity to desired accuracy, the matrix $\mathbf{M}_K$ is upper triangular to desired accuracy, and one can read off the Lyapunov exponents λ_a from the diagonal elements of the $\mathbf{R}_n(k)$'s:

$$\lambda_a = \frac{1}{2L} \sum_{j=1}^{2L} \log[R_K(j)_{aa}], \tag{5.48}$$

since the eigenvalues of a product of upper triangular matrices are the product of the eigenvalues of the individual matrices. The rapid rate of convergence of $\mathbf{Q}_K(2L)$ to the identity means one requires K no larger than two or three to see $\mathbf{Q}_K(2L)$ differing from the identity matrix by one part in 10^{-5} or 10^{-6}.

5.6.2 Smooth Coordinate Transformations

As we have noted local Lyapunov exponents $\lambda_i(\mathbf{x}, L)$ depend on the point on the attractor where the perturbation is initiated. This is in distinction to the global exponents which are independent of $\mathbf{x}$. Further, the global exponents are unchanged when we make a smooth change of coordinates on the phase space. The local exponents are not unchanged, but they acquire an additional piece which is only of order $\frac{1}{L}$. To see this, suppose we have created a state space of vectors $\mathbf{y}(k)$ by using time delay embedding or any other method of choice [LR91]. In the coordinates $\mathbf{y}$, the evolution of the system is

$$\mathbf{y}(n+1) = \mathbf{F}[\mathbf{y}(n)]. \tag{5.49}$$

Make a smooth (differentiable) nonlinear change in the coordinate system of the form

$$\mathbf{z} = \mathbf{H}(\mathbf{y}). \tag{5.50}$$

The mapping from time n to time n+1 in the z coordinates is given by

$$\begin{aligned} \mathbf{z}(n+1) &= \mathbf{G}(\mathbf{z}(n)) \\ &= \mathbf{H}(\mathbf{y}(n+1)) \\ &= \mathbf{H}(\mathbf{F}[\mathbf{y}(n)]) \\ &= \mathbf{H}(\mathbf{F}[\mathbf{H}^{-1}(\mathbf{z}(n))]). \end{aligned} \tag{5.51}$$

The Jacobian matrix $\mathbf{DG}(\mathbf{w})$ has elements

$$\begin{aligned} DG(\mathbf{w})_{\alpha\beta} &= \frac{\partial G_\alpha(\mathbf{w})}{\partial w_\beta} \\ &= DH(\mathbf{F}[\mathbf{H}^{-1}(\mathbf{w})])_{\alpha\gamma} DF(\mathbf{H}^{-1}(\mathbf{w}))_{\gamma\nu} DH^{-1}(\mathbf{w})_{\nu\beta}. \end{aligned} \tag{5.52}$$

When this is evaluated at $\mathbf{w} = \mathbf{z}(n)$, we find using $\mathbf{H}^{-1}(\mathbf{z}(n)) = \mathbf{y}(n)$ that

$$\mathbf{DG}(\mathbf{z}(n)) = \mathbf{DH}(\mathbf{y}(n+1)) \cdot \mathbf{DF}(\mathbf{y}(n)) \cdot [\mathbf{DH}(\mathbf{y}(n))]^{-1}. \tag{5.53}$$

In the last matrix in this product we used the fact that $\mathbf{H}[\mathbf{H}^{-1}(\mathbf{z})] = \mathbf{z}$, so

$$\mathbf{DH}^{-1}(\mathbf{z}(n)) = [\mathbf{DH}(\mathbf{y}(n))]^{-1}. \tag{5.54}$$

This now allows us to express the linearized dynamics of a perturbation $\delta\mathbf{z}(n)$ to the $\mathbf{z} \to \mathbf{G}(\mathbf{z})$ dynamics as

$$\delta\mathbf{z}(n+1) = \mathbf{DG}(\mathbf{z}(n)) \cdot \delta\mathbf{z}(n), \tag{5.55}$$

and to determine the perturbation L steps ahead of time n as

$$\delta \mathbf{z}(n+L) \;=\; \mathbf{DG}(n+L-1) \cdot \mathbf{DG}(n+L-2), \qquad (5.56)$$
$$\cdots, \mathbf{DG}(n)\delta \mathbf{z}(n)$$
$$=\; \mathbf{DG}^{L}(n)\delta \mathbf{z}(n),$$

where

$$\mathbf{DG}^{L}(n) = \mathbf{DH}(\mathbf{y}(n+L)) \cdot \mathbf{DF}^{L}(n) \cdot [\mathbf{DH}(\mathbf{y}(n))]^{-1}. \qquad (5.57)$$

Call the eigenvalues of the Oseledec matrix $\mathbf{OSL}(\mathbf{z}, L)$, $\kappa_i(\mathbf{z}, L)$, and note that

$$\frac{1}{2L} \log \det[\mathbf{OSL}(\mathbf{z}(n), L)] \;=\; \sum_{i=1}^{d} \kappa_i(\mathbf{z}(n),$$

$$=\; \sum_{i=1}^{d} \lambda_i(\mathbf{y}(n), L)$$

$$+\frac{1}{2L} \log \det\{\mathbf{DH}(\mathbf{y}(n+L)) \quad \cdot \quad [\mathbf{DH}(\mathbf{y}(n))]^{-1}\}. \qquad (5.58)$$

This means that the sum of local eigenvalues in the two coordinate systems differ by a term of order $\frac{1}{L}$. Similarly by considering all subdeterminants, one can demonstrate that the individual local exponents differ by order $\frac{1}{L}$. This is consistent with and actually constitutes the proof that the global exponents are unchanged under this smooth coordinate transformation. For that, one simply takes the limit $L \to \infty$ and uses the Oseledec theorem to remove the phase space dependence of the local exponents.

Basically the result here follows from the fact that the Jacobians in the $\mathbf{y}$ and in the $\mathbf{z}$ coordinate systems differ only by the terms involving $\mathbf{DH}$ at each end. These simply rescale and reorient the perturbation to the orbit, and in the limit of large L become unimportant to the exponential growth of the Jacobians composed L times along the orbit. The simple connection between the Jacobians is expected since we are observing the evolution in the tangent space to the attractor in each coordinate system. Because of the smoothness of the transformation $\mathbf{z} = \mathbf{H}(\mathbf{y})$, these tangent spaces are connected in a simple linear local fashion.

5.7   Local Lyapunov Exponents from Data

We already know how to estimate the local Jacobian of the dynamics from scalar data in a reconstructed time delay space, and the only twist required to find local Lyapunov exponents from the Oseledec matrix is to use the

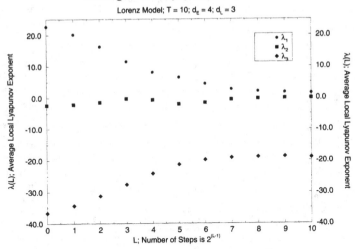

Average Local Lyapunov Exponent

Lorenz Model; T = 10; d_E = 4; d_L = 3

FIGURE 5.7.   The average local Lyapunov exponents $\bar{\lambda}_a(L)$ for data from the $x(t)$ dynamical variable of the Lorenz model, Equation (3.5), using $T = 10$ for time delay reconstruction of the data vectors in $d_E = 4$. We used local three-dimensional models $d_L = 3$ for the dimension of the local Jacobian matrices entering the determination of the $\bar{\lambda}_a(L)$. $d_L = 3$ was selected by local false nearest neighbors. We could have used $d_E = 3$, and the results would be essentially the same differing only at small L.

modified recursive QR decomposition since the times for which we wish its eigenvalues may well be short. To illustrate the results which come from this procedure we look at our standard examples.

5.7.1   Lorenz Model

We take our usual data set from the single variable $x(n)$ and form a d_E dimensional time delay state space. We know $d_E = d_L = 3$ from our earlier analysis of this system. From this we evaluate the local Jacobians as indicated above and using the recursive QR decomposition of the required product of Jacobians we compute the three $\bar{\lambda}_a(L)$ for the Lorenz system using $d_E = 4$, but of course we choose $d_L = 3$. These $\bar{\lambda}_a(L)$ are displayed in Figure 5.7. It is clear that one of the exponents is zero, one is positive, and the third is negative. The values we find for large L are $\lambda_1 = 1.51, \lambda_2 = 0.0$, and $\lambda_3 = -19.0$. With these values the Lyapunov dimension is found to be $D_L = 2.08$. In Figure 5.8 we examine these results in an expanded format. In the calculations just shown we used $d_E = 4$ for unfolding the attractor and a local dimension $d_L = 3$ for the dynamics. If we had used $d_E = 3$ there would have been little change. The message here is that using too large a d_E should cause no problem when one has clean data as it is d_L

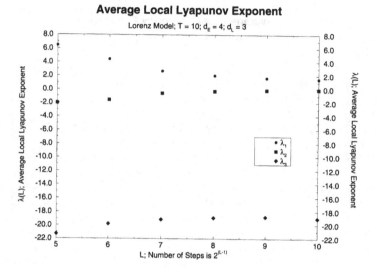

Average Local Lyapunov Exponent

FIGURE 5.8.   An expansion of Figure 5.7 with $5 \leq L \leq 10$ to illustrate the approach of the average local exponents to their global values λ_a. A zero global exponents is clear indicating that the data have come from a differential equation not a discrete time map.

which determines the number of true Lyapunov exponents. Differing d_E constitutes a change in coordinate system and alters the $O(1/L)$ term in our expression for the behavior of average local Lyapunov exponents. Having d_L correct is critical.

In Figure 5.9 we look at another aspect of local Lyapunov exponents by evaluating the exponents in $d_E = 4$ and $d_L = 4$. In the figure are shown both the four forward exponents and minus the four backward exponents. True exponents of the dynamics will change sign because of time reversal; however, false exponents associated with our having chosen too large a dynamical dimension d_E will behave otherwise. Indeed, we clearly see that three exponents change under reversing the way we read the data in time, and one does not. Unfortunately this way of determining d_L is not robust against contamination.

5.7.2   Nonlinear Circuit with Hysteresis

In analyzing the local Lyapunov exponents from this data source we have our choice of two measured quantities. If we use V_A we must work in a global embedding space $d_E = 5$ and make local three-dimensional mappings from neighborhood to neighborhood. This works, but not as accurately as using V_B and making these same mappings with a global three-dimensional phase space. Basically when we have any contamination on our data the evaluation of Lyapunov exponents can be difficult. If our attractor has

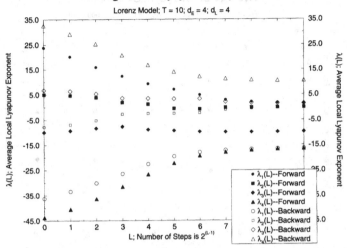

Average Local Lyapunov Exponent

FIGURE 5.9.   The average local Lyapunov exponents $\bar{\lambda}(L)_a$ for data from the $x(t)$ dynamical variable of the Lorenz model, Equation (3.5), using $T = 10$ for time delay reconstruction of the data vectors in $d_E = 4$. We used local three-dimensional models $d_L = 4$ for the dimension of the local Jacobian matrices entering the determination of the $\bar{\lambda}_a(L)$, so one of the Lyapunov exponents is false. The data are presented for evaluation of $\bar{\lambda}_a(L)$ forward and (minus) backward in time along the attractor. The equality of three of the $\bar{\lambda}_a(L)$ forward and (minus) backward means $d_L = 3$ which we have also established with local false nearest neighbors.

$d_L = 3$ and can be embedded in $d_E = 3$ rather than $d_E = 5$, then we do not populate the other two dimensions with contamination and can more accurately determine each of the required local Jacobians. One can demonstrate with simulated data where as much uncontaminated data can be made available, that working in $d_E > d_L$ can give d_L accurate exponents.

So we use data from V_B confident that the same numerical values would be achieved by using $d_E = 5$ and $d_L = 3$ for V_A data. The local Lyapunov exponents which result are displayed in Figure 5.10. The appearance of a zero exponent tells us we have differential equations underlying this data, and that is what we expect. The positive exponent is $\lambda_1 = 0.26$ and the third exponents is $\lambda_3 = -0.56$. From these values we determine a Lyapunov dimension $D_L = 2.51$. We should be able to predict this circuit forward in time about

$$\frac{\tau_s}{\lambda_1} \approx 0.4 \text{ ms.} \tag{5.59}$$

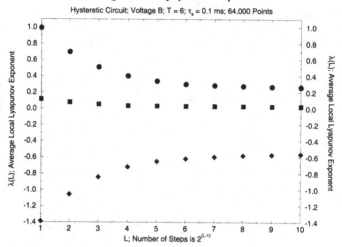

Average Local Lyapunov Exponent

Hysteretic Circuit; Voltage B; T = 6; τ_s = 0.1 ms; 64,000 Points

FIGURE 5.10.   The average local Lyapunov exponents $\bar{\lambda}(L)_a$ for data from the $V_B(t)$ dynamical variable of the nonlinear circuit with hysteretic element using $T = 6$ for time delay reconstruction of the data vectors in $d_E = 3$. We used local three dimensional models $d_L = 3$ for the dimension of the local Jacobian matrices entering the determination of the $\bar{\lambda}_a(L)$. $d_L = 3$ was selected by local false nearest neighbors. Sixty four thousand data points were used here.

5.8 A Few Remarks About Lyapunov Exponents

As classifiers of the source of a chaotic signal, Lyapunov exponents carry a substantial advantage over the use of fractal dimensions. Lyapunov exponents, especially local exponents, not only bring a set of numbers for classifying the system but also tell us the limits to predictability of the chaotic system. The Lyapunov dimension also gives one a sense of dimension which corresponds accurately to other fractal dimension estimates. It seems that the combination of fractal dimensions, when they can be accurately estimated, or better, whole curves of the correlation functions $C(q, r)$ nicely compliment the Lyapunov exponents as classifiers for the dynamics. Estimating all d_L Lyapunov exponents for a system is quite time consuming if the local exponents are desired as well, but this is a limitation which is likely to be overcome by parallelizing the computations. Doing these computations in serial mode is not natural. Really one wants to estimate the local Jacobian matrices $\mathbf{DF}(\mathbf{y}(k))$ which enter the Oseledec matrix. This can be done at many phase space locations at the same time. One can expect a speed up in this kind of computation which is linear in the number of processors. In the near future one should be able to handle large data sets and large d_L with this kind of improvement.

Our ability to estimate the negative Lyapunov exponents requires a comment, for it may seem that with data on the attractor alone we should not be able to determine Lyapunov exponents which govern the way orbits move to the attractor from regions in the basin of attraction surrounding the attractor. The key is that the vector field which determines the evolution of orbits is analytic in its arguments. When we make a local neighborhood to neighborhood map we are determining the local structure of the d_L-dimensional vector field including those aspects of it which give rise to negative Lyapunov exponents. Our ability to do this comes from the use of neighborhoods near points on the attractor. This effectively explores regions around attractor points and gives a local vector field which operates in d_L dimensions and not on the attractor alone.

A part of this general subject which is of some importance in applications and requires additional attention is the description of local exponents $\lambda_a(\mathbf{x}, L)$ as a function of phase space location $\mathbf{x}$ as well as a nice means for locating those regions where all the $\lambda_a(\mathbf{x}, L)$ are negative. Such regions stand out as the phase space places to which one might want to control a system because the local stability is so marked. Another important topic is that of exploring the idea of regional Lyapunov exponents which involve averaging over real space locations in the underlying system. We will see the role this might play when we look at data from the Great Salt Lake and from the boundary layer chaos in Section 11.2. In each case we are seeing data with small scale motions suppressed yet with very interesting dynamics on the larger scales sensed in the observations. How the Lyapunov exponents of a system vary with spatial averaging is the issue.

6
Modeling Chaos

6.1 Model Making in Chaos

We next discuss making models for prediction or control of the source of the observed chaotic signal. In a sense this is both the easiest and the hardest task we have discussed. It is the easiest because it is quite simple to make models of the dynamics which very accurately allow one to predict forward in time from any new initial condition close to or on the attractor within the limits of the intrinsic instabilities embodied in the positive Lyapunov exponents. It is also the hardest because there is no guideline as to which of many functional forms to use for the models and what interpretation to place on the parameters in the models from a physical point of view. In this section we make models on the attractor and evaluate them by how well they do in prediction or possibly prediction of Lyapunov exponents.

Another route to model making is based in an understanding of the fundamental physics or biology, etc, of the problem. One develops equations of motion for the dynamical system and then compares the output of those equations of motion to properties of the observations $\mathbf{y}(k)$. This comparison is not by individual computed orbits $\mathbf{y}(k)$ to observed orbits $\mathbf{y}(k)$, for these must disagree and be essentially uncorrelated from each other again due to the intrinsic instabilities in the dynamics. Instead the comparison is to be made in terms of the statistical quantities such as fractal dimensions and Lyapunov exponents as we have discussed.

We work with data as observed on the attractor alone, thus we cannot hope as an algorithmic matter to make models which would have general validity throughout the system state space. For example, if there is a set of initial conditions in that original phase space which leads to other behavior than what we have observed and analyzed, it is plausible that our models for the motion in that other basin of attraction could be different. It is also quite possible that we would have been lucky enough or insightful enough to have made a model which encompasses both features of the dynamics. When we make models we have to decide from the outset what kind of functional form we are going to use to express what is certainly fundamentally forms of Newton's laws. The impossibility of selecting in any *a priori* way the correct functional form is stressed by Rissanen [Ris89] who also lays out a clear program for extracting from experimental data the most information available even though the functional form is not specified.

One can use the phase space structure we have built up in the $\mathbf{y}(n)$ to provide effective models of the dynamics which allow one to predict the evolution of any new point in the phase space within the basin of attraction which has been observed. The basic idea is that since we have seen how points in a neighborhood evolve into points in the "next" neighborhood we ought to be able to provide an appropriate interpolation scheme which would allow us to say that any new point would evolve more or less as its neighborhood was seen to evolve.

If the $\mathbf{y}(n) = [s(n), s(n + T), \ldots, s(n + (d_E - 1)T)]$ are made out of observations $s(1), s(2), \ldots$, then our predictive models will only be able to provide predictions for future values of that variable. If we wish to provide a predictive model for any other dynamical variable $v(t)$, then

a we must have dynamical equations connecting $s(t)$ and $v(t)$, or

b we must have simultaneously measured $s(t)$ and $v(t)$ over some period in the past. Our knowledge of the reconstructed phase space of the $\mathbf{y}(n)$ would then allow us to build a dynamical connection $v(n) = G(\mathbf{y}(n))$ in that phase space. We will take this up in Section 6.4.

6.2 Local Models

We first consider local models which consist of local neighborhood to neighborhood maps in the reconstructed phase space. The idea follows our construction of local maps to extract the local Jacobians used in determining Lyapunov exponents. We start with a specified local functional form for the dynamics $\mathbf{x} \to \mathbf{F}^{(k)}(\mathbf{x})$ in the neighborhood of the observed point $\mathbf{y}(k)$:

$$\mathbf{F}(\mathbf{x}, k) = \sum_{m=1}^{M} \mathbf{c}(m, k)\phi_m(\mathbf{x}), \qquad (6.1)$$

where the functions $\phi_m(\mathbf{x})$ are a basis set we choose from intuition or good guessing or convenience. These $\phi_m(\mathbf{x})$ could be polynomials or other functions with some appeal. The discussion of what functions to use and how many to use is the subject of multidimensional interpolation [Pow81]. In an intuitive sense we can say that if we have enough data, then local polynomial approximations to the dynamics are sure to provide accurate local maps. When data becomes sparse or dimensions become high and the number of coefficients in the polynomials correspondingly large, other interpolation functions will probably be more efficient and accurate. Radial basis functions offer an attractive choice, and we shall consider them below.

Return now to the general problem, which we then illustrate with polynomials. We go to a point $\mathbf{y}(k)$ in the embedding space of dimension d_E and using a device such as the principal component decomposition discussed in the context of local false nearest neighbors, select out a d_L-dimensional subspace in which to make a model. All distances are evaluated in the $d_E \geq d_L$ dimensional space, but all other computations are done on the d_L dimensional model which takes the selected d_L components of $\mathbf{y}(k)$ into the same d_L components of $\mathbf{y}(k+1)$ via

$$
\begin{aligned}
\mathbf{y}(k+1) &= \mathbf{F}(\mathbf{y}(k), k) \\
&= \sum_{m=1}^{M} \mathbf{c}(m,k)\phi_m(\mathbf{y}(k)).
\end{aligned} \tag{6.2}
$$

To determine the coefficients in the model we locate the N_B nearest neighbors $\mathbf{y}^{(r)}(k); r = 1, 2, \ldots, N_B$, of $\mathbf{y}(k)$ and minimize

$$
\sum_{r=1}^{N_B} |\mathbf{y}(r; k+1) - \sum_{m=1}^{M} \mathbf{c}(m,k)\phi_m(\mathbf{y}^{(r)}(k))|^2. \tag{6.3}
$$

This is a linear problem once the basis functions $\phi_m(\mathbf{x})$ are fixed. Vary this with respect to the $c_\beta(n,k); \beta = 1, 2, \ldots, d_L$, and find

$$
\sum_{m=1}^{M} M(k)_{nm} c_\beta(m,k) = \sum_{r=1}^{N_B} y_\beta(r, k+1)\phi_n(y^{(r)}(k)), \tag{6.4}
$$

where

$$
M(k)_{mn} = \sum_{r=1}^{N_B} \phi_n(\mathbf{y}^{(r)}(k))\phi_m(\mathbf{y}^{(r)}(k)). \tag{6.5}
$$

So the problem is a standard $M \times M$ matrix inversion problem

$$
c_\beta(m,k) = \sum_{n=1}^{M} (M(k)^{-1})_{mn} \left[\sum_{r=1}^{N_B} y_\beta(r, k+1)\phi_n(y^{(r)}(k)) \right], \tag{6.6}
$$

and this is well studied. There are many excellent algorithms available [GV89]. When we have determined the $c_\beta(m, k); \beta = 1, 2, \ldots, d_L; m = 1, 2, \ldots, M$ for each $k = 1, 2, \ldots, N$ in the data set, these numbers provide a lookup table for interpolation methods on and near the attractor. We will have a local model associated with each observed point $\mathbf{y}(k)$ on the attractor.

In practice to predict ahead from a new point $\mathbf{z}(0)$ we search through the $\mathbf{y}(k)$ to find the one nearest $\mathbf{z}(0)$; call it $\mathbf{y}(J)$. We now lookup the model local to $\mathbf{y}(J)$. This is $\mathbf{F}(\mathbf{x}, J)$, and it should be valid as an interpolating function in the neighborhood of $\mathbf{y}(J)$ and $\mathbf{z}(0)$. Next evaluate $\mathbf{F}(\mathbf{z}(0), J)$, and this gives us the next point on the orbit which starts with $\mathbf{z}(0)$ as initial condition: $\mathbf{z}(1) = \mathbf{F}(\mathbf{z}(0), J)$. We call this an interpolating operation because the function $\mathbf{F}(\mathbf{x}, J)$ contains information in its coefficients $\mathbf{c}(m, J)$ about all the neighbors in the neighborhood of $\mathbf{y}(J)$ and provides information throughout that neighborhood in a locally smooth fashion. Next find the nearest neighbor of $\mathbf{z}(1)$, call it $\mathbf{y}(K)$, and lookup the required local map $\mathbf{F}(\mathbf{x}, K)$ to proceed to $\mathbf{z}(2) = \mathbf{F}(\mathbf{z}(1), K)$. Iterate this procedure as far into the future of $\mathbf{z}(0)$ as desired. The bound on our accuracy in this is determined by the error we make in the actual value of $\mathbf{z}(0)$ and the largest local Lyapunov exponents $\lambda_1(\mathbf{z}(l), L)$ which tells us how that error grows locally near each $\mathbf{z}(l)$.

The procedure we have just described is called iterative forecasting since we make a large number of unit time steps to reach L steps into the future of $\mathbf{z}(0)$. The root mean square error in this forecast should scale approximately as [ABST93, FS89] in going from $\mathbf{z}(0)$ to $\mathbf{z}(L)$ in L steps

$$N^{-\frac{(P+1)}{d_L}} e^{L\lambda_1}, \tag{6.7}$$

if we use polynomial basis functions $\phi_n(\mathbf{x})$ where P is the maximum order of the polynomials used, and N is the number of data.

If we build a model which goes from $\mathbf{z}(0)$ to $\mathbf{z}(L)$ in one direct step, then the scaling is less optimistic with the RMS error estimated to be

$$N^{-\frac{(P+1)}{d_L}} e^{(P+1)L\lambda_1} \tag{6.8}$$

after L steps. This direct prediction across L steps requires much more accurate interpolation methods as the neighborhood of phase space we must deal with is much larger. By proceeding in smaller steps, as in the iterative predictions, we can correct along the way for errors which inevitably creep into our predictions.

6.2.1 Lorenz Model

As ever we will begin with an example from the Lorenz model to demonstrate how these methods work. In Figure 6.1 we have the RMS prediction

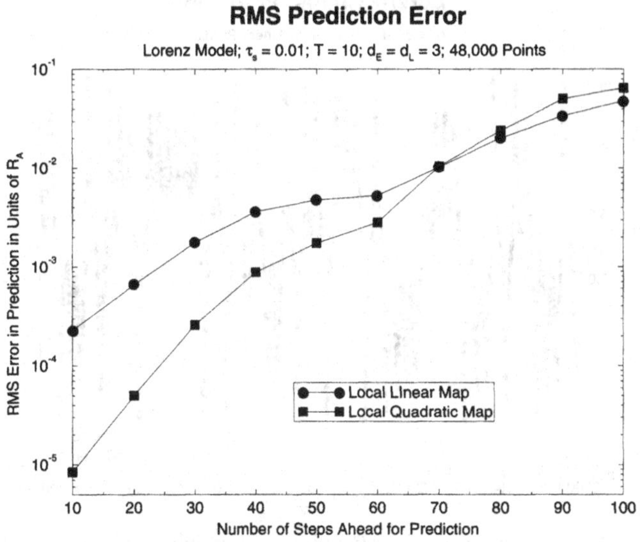

RMS Prediction Error

Lorenz Model; τ_s = 0.01; T = 10; d_E = d_L = 3; 48,000 Points

FIGURE 6.1.   The RMS prediction error for local linear and local quadratic map models for data from $x(t)$ from the Lorenz system, Equation (3.5). Models were built in $d_E = d_L = 3$ using $T = 10$ as determined using average mutual information. The error is in units of the size of the attractor R_A. The error grows approximately exponentially at a rate dictated by the largest Lyapunov exponent λ_1.

error scaled to the size of the attractor R_A for local polynomial prediction functions. The results for local linear maps are shown with circles and local quadratic maps, with squares. The error grows approximately exponentially with the number of steps ahead of any given point. The computation was done by using 48,000 data points in a reconstructed phase space $d_E = 3$ and local maps with $d_L = 3$. One thousand different initial conditions were examined, and we have displayed the average over these starting locations. A specific prediction for the Lorenz system is seen in Figure 6.2 Thirty thousand points from the $x(t)$ time series for the Lorenz attractor were used to make three dimensional models ($d_L = 3$) in a space of three dimensions ($d_E = 3$). These were local linear models. Using these models the evolution of the Lorenz system was predicted 25 steps ahead of τ_s in a recursive fashion. That is to say, the models made predictions one step at a time and we iterated these maps 25 times to arrive at the predictions shown in Figure 6.2 where points 35,000 through 37,500 are predicted with striking accuracy. Twenty five steps of τ_s represents a growth of any initial error by $e^{25\tau_s \lambda_1} \approx 12.2$ using $\tau_s = 0.01$ and $\lambda_1 = 1.51$, and this probably represents the limit of good predictability on the average. Locally the predictions may be much better or much worse depending on $\lambda_a(\mathbf{x}, L)$.

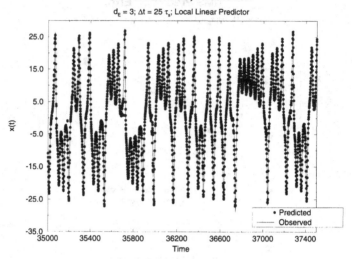

FIGURE 6.2.   Observed and predicted $x(t)$ dynamical variable for the Lorenz model, Equation (3.5). The predictions were made using local linear polynomial predictors whose coefficients were learned from 30,000 points of data from $x(t)$. These were embedded in $d_E = 3$ using a $d_L = 3$ dimensional model. Predictions were made 1 step at a time for 25 steps.

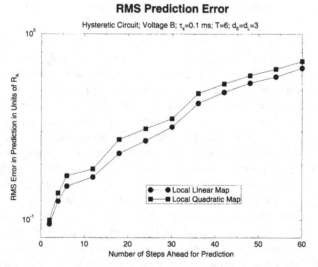

FIGURE 6.3.   The RMS prediction error for local linear and local quadratic map models for data from $V_B(n)$ from the hysteretic nonlinear circuit. Models were built in $d_E = d_L = 3$ using $T = 6$ as determined using average mutual information. The error is in units of the size of the attractor R_A. The error grows approximately exponentially at a rate dictated by the largest Lyapunov exponent λ_1.

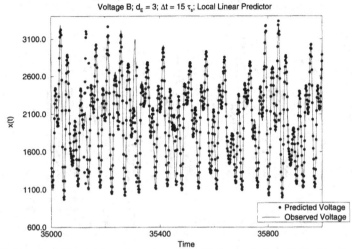

FIGURE 6.4.   Observed and predicted $V_B(t)$ dynamical variable for the nonlinear circuit with hysteretic element. The predictions were made using local linear polynomial predictors whose coefficients were learned from 30,000 points of data from $V_B(t)$. These were embedded in $d_E = 3$ using a $d_L = 3$ dimensional model. Predictions were made 1 step at a time for 15 steps.

6.2.2   Nonlinear Circuit with Hysteresis

In Figure 6.3 we forecast using data from the hysteretic circuit discussed earlier. The data was taken from the voltage V_B, and 48,000 data points in $d_E = 3$ were used to make local linear and then local quadratic maps in $d_L = 3$. The average RMS error relative to the size of the attractor is shown in the figure with an average over 1000 starting sites.

Using local linear polynomial maps in $d_E = d_L = 3$ we predicted ahead for times of 15 then 25 $\tau_s = 0.1$ ms for voltage B of the hysteretic nonlinear circuit. Thirty thousand points from the experiments of Carroll and Pecora were used to learn the local maps, and then predictions were made using the methods we have described. In Figure 6.4 we have the predictions compared to the observed voltages for 15 steps ahead from any new point on the attractor. In Figure 6.5 we see the same results for predicting 25 steps ahead. In each case we used the local models derived from the first 30,000 data points to predict from points 35,000 to 36,000 as shown. The degradation of the predictability is clear as we extend ourselves further in time, and this is as it must be for chaos. There are regions of enormously accurate predictability and regions with much lower predictability. This too is consistent with the large variations of local Lyapunov exponents we have seen in model systems.

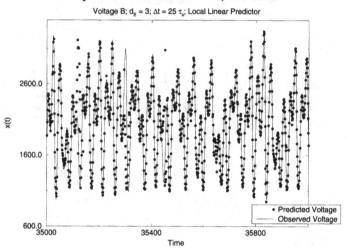

FIGURE 6.5.   Observed and predicted $V_B(t)$ dynamical variable for the nonlinear circuit with hysteretic element. The predictions were made using local linear polynomial predictors whose coefficients were learned from 30,000 points of data from $V_B(t)$. These were embedded in $d_E = 3$ using a $d_L = 3$ dimensional model. Predictions were made 1 step at a time for 25 steps. The quality of the predictions here is markedly worse than those in Figure 6.4.

6.2.3   A Few Remarks About Local Models

Making accurate local models on the attractor using neighborhood to neighborhood information has proven rather an easy task. The local maps are determined by coefficients of a selected basis set which carries the phase space information. The evaluation of the coefficients by requiring a best least-squares fit to the evolution of whole neighborhoods of points around the observed orbit is a thoroughly studied linear matrix inversion problem. We will exhibit the use of this method on field and laboratory data in a later chapter, and we will show it to be quite useful in a practical sense. Our computations were confined to the use of local polynomials in the neighborhoods of phase space, but we could have done as well or better by using more sophisticated functions [Pow81].

The local method has limitations which include the need to have local information to proceed at all. This typically means that the data set must populate the whole attractor rather well so the neighborhoods are also well populated. To achieve accuracy in the determination of the coefficients in the local maps, we usually require twice the number of neighbors as the number of local coefficients. When data is sparse, this may be too heavy a demand, and it may be necessary to turn to global models of the form we soon describe.

When data is plentiful, local models can work extremely well. One should not be led to conclude that their success in the sense of an elaborate, but useful, lookup table for phase space interpolation provides any sense of the dynamics of the source. When parameters of the source change, no rule can be extracted from these local models, they require recalibration as the attractor may well change under the parameter change. Further, from local rules on an attractor of some system, we are unlikely to be able to infer a global dynamical rule which would apply well off the attractor or in another part of state space where a quite different attractor may be operative.

6.3 Global Models

The full collection of local maps form a model which is useful over the whole attractor though it is unlikely to be of value off the attractor. Its discontinuities from neighborhood to neighborhood and its extremely large number of adjustable parameters are also shortcomings of local models. For polynomial models of order P in d_L local dimensions we have approximately d_L^P parameters at each time step. This is clearly a penalty for high accuracy. It would be economical to have relatively simple continuous model describing the whole collection of data. We suspect there is such a model underlying the data anyway, but guessing the functional form from information about the attractor alone is difficult—maybe not possible. Nonetheless, a number of global models have been discussed which present a closed functional representation of the dynamics in the phase space on and near the whole attractor. The smoothness or analyticity in phase space of such global models as well as the ability to produce orbits over large regions of the phase space makes them quite attractive. Global models, when they work well, also provide an efficient framework in which to examine changes in system parameters.

Global modeling also uses an expansion of the $\mathbf{x} \rightarrow \mathbf{F}(\mathbf{x})$ vector field $\mathbf{F}(\mathbf{x})$ in a selected set of basis functions in d_E dimensions. This is essentially what we used for local models, but now one applies the idea over the whole attractor at once. A natural global method is to use polynomials again. There is an attractive approach to finding a polynomial representation of a global map. This *measure-based functional reconstruction* [GLC91, Bro93] uses orthogonal polynomials whose weights are determined by the invariant density on the attractor. This method eliminates the problem of multiparameter optimization. Finding the coefficients of the polynomials and the coefficients of the function $\mathbf{F}(\mathbf{x})$ requires only the evaluation of moments of data points in phase space. The method works as follows. Introduce polynomials $\phi_m(\mathbf{x})$ on R^{d_E} which are orthogonal with respect to the natural invariant density

$$\rho(\mathbf{x}) = \frac{1}{N} \sum_{k=1}^{N} \delta^{d_E}(\mathbf{x} - \mathbf{y}(k)) \tag{6.9}$$

on the attractor. This orthogonality requires

$$\int d^{d_E}x \, \rho(\mathbf{x}) \phi_m(\mathbf{x}) \phi_n(\mathbf{x}) = \delta_{mn}. \tag{6.10}$$

The polynomials are determined by a conventional Gram-Schmidt procedure starting from

$$\phi_1(\mathbf{x}) = 1, \tag{6.11}$$

and determining the other polynomials up to the order selected.

The global vector field $\mathbf{F}(\mathbf{x})$ which evolves data points as $\mathbf{y}(k+1) = \mathbf{F}(\mathbf{y}(k))$ is approximated in M^{th} order as

$$\mathbf{F}_M(\mathbf{x}) = \sum_{m=1}^{M} \mathbf{c}(m) \phi_m(\mathbf{x}). \tag{6.12}$$

This differs from the local expansion of the same form by having the coefficients independent of the phase space location where the approximation is being made. It is global. The coefficients $\mathbf{c}(m)$ are determined via

$$
\begin{aligned}
\mathbf{c}(m) &= \int d^{d_E}x \, \mathbf{F}(\mathbf{x}) \phi_m(\mathbf{x}) \rho(\mathbf{x}), \\
&= \frac{1}{N} \sum_{k=1}^{N} \mathbf{F}(\mathbf{y}(k)) \phi_m(\mathbf{y}(k)), \\
&= \frac{1}{N} \sum_{k=1}^{N} \mathbf{y}(k+1) \phi_m(\mathbf{y}(k)), \tag{6.13}
\end{aligned}
$$

using

$$\mathbf{y}(k+1) = \mathbf{F}_M(\mathbf{y}(k)). \tag{6.14}$$

This demonstrates the power of the method directly. Once the orthogonal polynomials $\phi_m(\mathbf{x})$ are determined from the data, the evaluation of the vector field is reduced to sums over powers of the data with themselves since the $\phi_m(\mathbf{x})$ are polynomials.

The form of the sums involved allow one to establish the vector field from a given set of data and adaptively improve it as new data are measured. The best aspect of the method, however, may be the robustness against contamination of the data [Bro93]. There is no least-squares parameter search involved, so no distances in state space need be evaluated. The geometric nature of the method doesn't rely on accurately determining distances and is thus not so sensitive to "noise" which spoils such distance evaluations. By using the whole data set instead of just data in a local neighborhood a certain amount of averaging and thus "noise" filtering is done automatically.

6.3.1   The Ikeda Map

We give an example of this global modeling method by considering its application to the Ikeda map [Ike79, HJM85]. The global model is then used to evaluate global Lyapunov exponents by extracting the Jacobian matrices along an orbit of the system. We recall the Ikeda map for the point $z(n) = x(n) + iy(n)$ in the complex plane

$$z(n+1) = p + Bz(n)\exp[i\kappa - i\alpha/(1 + |z(n)|^2)] \qquad (6.15)$$

where we will use the parameters $p = 1.0$, $B = 0.76$, $\kappa = 0.4$, and $\alpha = 6.0$ for this discussion. The dimension of the attractor associated with this map is $d_A \approx 1.4$, and the use of global false nearest neighbors shows that $x(n)$ data from this system can be embedded in $d_E = 3$. Note that the parameters for the Ikeda map are different here than those used in Section 4.7. We use this information to make a global model utilizing $M = 10$ orthogonal polynomials according to the description above. The Jacobian matrix for the dynamics, which here has a global functional form, is evaluated along the trajectory and used in the Oseledec matrix to yield Lyapunov exponents. One of the exponents is false and is identified by its behavior under time reversal. The other two exponents are approximately $\lambda_1 \approx 0.35$ and $\lambda_2 \approx -1.0$.

The $x(n)$ data from the map is next contaminated with Gaussian noise from a random number generator, and the evaluation of the Lyapunov exponents is performed with varying levels of noise. Figure 6.6 displays the two true Lyapunov exponents as a function of the amplitude of the noise level. Eleven hundred data points from the map are used to create the global polynomial map. The RMS level of the Ikeda map is about 0.7. The largest contamination level has an RMS level of 0.1 meaning about a 14 percent noise level in amplitude. In decibels (dB) this translates into a signal to noise ratio of 16.9 dB. The signal to noise ratio when the signal is cleanest, having noise amplitude level 0.001, is 56.9 dB. The results of using the global map to evaluate the Lyapunov exponents are shown in solid symbols. Using the local polynomial method to evaluate Lyapunov exponents from the $x(n)$ data embedded in $d_E = 3$ gives the output shown in open symbols. It is clear that the global polynomial method, which requires substantially less computation, is much more robust against noise contamination.

6.3.2   Other Global Methods

Another kind of nonlinear modeling combines features of local and global models. An example is the method of *radial basis functions* [Par92, Pow81] which [Cas89], "is a global interpolation technique with good localization

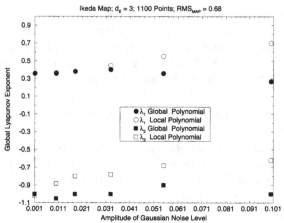

FIGURE 6.6.   The global Lyapunov exponents for the Ikeda map of the plane to itself. The exponents are displayed as a function of the amplitude of Gaussian random noise added to the "observed" signal from the map. The RMS size of the signal is 0.7 in these units. The results in solid symbols comes from using a global orthogonal polynomial representation of the mapping $\mathbf{x} \to \mathbf{F}(\mathbf{x})$; the data in open circles comes from making local polynomial neighborhood to neighborhood maps. The global polynomial method requires less computation than local models and is often more robust against contamination of the data.

properties." In this method a predictor $\mathbf{F}(\mathbf{y})$ is sought in the form

$$\mathbf{F}(\mathbf{y}) = \sum_{n=1}^{N_c} \mathbf{c}(n)\Phi(\|\mathbf{y} - \mathbf{y}(n)\|) \tag{6.16}$$

where $\Phi(\|\mathbf{x}\|)$ is some smooth function. The coefficients $\mathbf{c}(n)$ are chosen to minimize the residuals in the usual least-squares fit to the data. Depending on the number of points N_c used for reconstruction this method can be considered as local when $N_c \ll N$ or global when $N_c \approx N$.

Various choices for $\Phi(\|\mathbf{x}\|)$ will do. $\Phi(r) = (r^2 + c^2)^{-\beta}$ works well for $\beta > -1$ and $\beta \neq 0$. If one adds a sum of polynomials to the sum of radial basis functions, then even increasing functions $\Phi(r)$ provide good localization properties. However, for a large number of points N_c this method is as computationally expensive as the usual least-square fit. Numerical experiments carried out in [Cas89] show that for small number of data points radial basis predictors do a better job than polynomial models. For larger amounts of data local polynomial models seem to be superior.

A interesting variant of radial basis functions is kernel density estimation [Sil86]. In this method one estimates a smooth probability distribution from discrete data points. Each point is associated with its *kernel* which is a smooth function $K(\|\mathbf{y} - \mathbf{y}(i)\|)$ which typically decays with distance, but sometimes can even increase. Using a kernel chosen *a priori*, a probability distribution

$$p(\mathbf{x}) = \sum_i K(\|\mathbf{x} - \mathbf{y}(i)\|) \tag{6.17}$$

or a conditional probability distribution

$$p_c(\mathbf{x}|\mathbf{z}) = \sum_i K(\|\mathbf{x} - \mathbf{y}(i+1)\|)\, K(\|\mathbf{z} - \mathbf{y}(i)\|) \tag{6.18}$$

is estimated. $p_c(\mathbf{x}|\mathbf{z})$ can then be used for conditional forecasting by the rule for the estimated phase space point evolving from $\mathbf{y}(k)$

$$\tilde{\mathbf{y}}(k+1) = \int d\mathbf{x}\, \mathbf{x} p_c(\mathbf{x}|\mathbf{y}(k)). \tag{6.19}$$

Kernel density estimation usually provides the same accuracy as the first-order local predictors.

In computing the conditional probability distribution, one can impose weights in order to attach more value to the points close to the starting point of prediction (both in time and in phase space). In fact, this leads to a class of models which are hybrids of local and global methods. Moreover, it allows one to construct a model which possesses not only good predicting properties but also preserves important invariants of the dynamics. The prediction model by Abarbanel, Brown, and Kadtke [ABK90] belongs to this class. That model chooses the global map as

$$\mathbf{F}(\mathbf{y}, \mathbf{a}) = \sum_{k=1}^{N-1} \mathbf{y}(k+1)g(\mathbf{y}, \mathbf{y}(k); \mathbf{a}), \tag{6.20}$$

where $g(\mathbf{y}, \mathbf{y}(k); \mathbf{a})$ is the analog of the kernel function. It is near one for $\mathbf{y} = \mathbf{y}(k)$ and vanishes rapidly away from there. The $\mathbf{a}$ are constants. To determine the $\mathbf{a}$ and additional parameters $\mathbf{X}$ minimize

$$\sum_{k=L}^{N-1} \left\| \mathbf{y}(k+1) - \sum_{n=1}^{L} X_n \mathbf{F}^n(\mathbf{y}(k-n+1), \mathbf{a}) \right\|^2, \tag{6.21}$$

where $X_n, n = 1, .., L$ is a set of weights attached to the sequence of points $\mathbf{y}(k-n+1)$ *all* of which are to be mapped into $\mathbf{y}(k+1)$ by maps $\mathbf{F}^n(\mathbf{y})$. It is clear that then $\mathbf{F}(\mathbf{y}(k), \mathbf{a})$ will be close to $\mathbf{y}(k+1)$ as it provides an excellent interpolation function in phase space and in time. The free parameters in the kernel function g and the specific choice of the cost function allow one to predict forward accurately in time *and* satisfy additional constraints imposed by the dynamics—significant Lyapunov exponents and moments of the natural invariant density distribution as determined by the data will be reproduced in this model. The latter does not follow automatically from the former. Indeed, it is shown in [ABK90] that global models which predict enormously accurately can have all negative Lyapunov exponents even when the data are chaotic. The orthogonal polynomial models explored in [GLC91, Bro93] seem to circumvent this problem at least when applied to simple models.

It should be clear that each of these modeling methods relies on local neighborhood information to provide the weighting functions, and all are an interesting variant of the use of phase space information to provide temporal predictive power.

6.3.3 A Few Remarks about Global Model Making

Global models of the source of one's observations are in some sense the eventual goal of the analysis of any data set. Newton's equations certainly describe much more than the Keplerian orbits which suggested them. The development of such global models is unlikely to proceed in an algorithmic sense. Rissanen [Ris89] discusses model making from data in this general kind of context and suggests that since the number of functional forms available for global models is infinite, no algorithmic rule for selection will ever work.

The fact that the global model one builds may not be "the correct" model does not diminish the usefulness of global models when they exist for the problem at hand. Using global polynomials everywhere is unlikely to provide workable global models for all situations as the vector fields can have quite different global analyticity properties than possessed by simple polynomials. In the region of phase space occupied by the attractor on which one's data is located these differences may be trivial. If the vector field is composed of trigonometric functions and we are making polynomial approximations, something must eventually give. Again global models may be useful for answering many questions, but without a sense of the fundamental rules for the physics or chemistry or biology or other scientific basis of the source, they are as much elaborate bookkeeping devices as the local models discussed earlier.

6.4 Phase Space Models for Dependent Dynamical Variables

In the description of dynamical models using known variables and their derivatives, once we have acquired enough coordinates to fully capture the motion, we can, in principle, provide a functional form for any other variables of the system in terms of the ones we have selected. Basically once we have an exhaustive set of variables which describes the motion of the system, all others are redundant.

This provides an interesting feature of time delay reconstruction or, for that matter, any reconstruction of phase space using scalar variables and simple actions on them [ACP*94]. Suppose we observe two or more variables of some dynamical system, starting with $s_A(t)$ and $s_B(t)$. Now choose to reconstruct phase space with $s_A(t)$ and its time delay variables chosen as we have discussed above. So from $s_A(n) = s_A(t_0 + n\tau_s)$ we form

$$\mathbf{y}_A(n) = [s_A(n), s_A(n+T), \ldots, s_A(n + (d_E - 1)T)]. \tag{6.22}$$

Since this provides a full set of coordinates for the observations, essentially by the definition of d_E, it should be possible to create a local model for the behavior of $s_B(n)$ using $d_L \le d_E$ of the coordinates in the $\mathbf{y}_A(n)$. Call the d_L-dimensional vectors made from $\mathbf{y}_A(n)$, $\mathbf{Y}_A(n)$. This local model will relate $s_B(n)$ to the $\mathbf{Y}_A(n)$ locally in reconstructed phase space

$$s_B(n + m) = F_n(\mathbf{Y}_A(n)), \tag{6.23}$$

where $m\tau_s$ is a possible delay between the time delay vector $\mathbf{Y}_A(n)$ and the observations $s_B(n)$. In many cases $m = 0$ may do. The local function $F_n(\bullet)$ could be a polynomial or other more complex function of its arguments. It will relate vectors $\mathbf{Y}_A(n)$ and their neighbors $\mathbf{Y}_A^{(r)}(n)$ to the measured values of $s_B^{(r)}(n)$ corresponding to the appropriate observation times of those neighbors. The neighbors of $\mathbf{Y}_A(n)$ are to be determined in $d_E \ge d_L$ dimensional space as that is where the false nearest neighbor test has established that all neighbors are true neighbors. The d_L-dimensional subspace of this unfolding space is used for local model making only.

To be more explicit in the case of functions made from local linear polynomials, we would write

$$F_n(\mathbf{Y}_A(n)) = A(n) + \mathbf{B}(n) \cdot \mathbf{Y}_A(n), \tag{6.24}$$

and determine the components of the scalars $A(n)$ and the vectors $\mathbf{B}(n)$ as usual from minimizing

$$\sum_{r=0}^{N_B} |s_B^{(r)}(n) - A(n) - \mathbf{B}(n) \cdot \mathbf{Y}_A^{(r)}(n)|^2, \tag{6.25}$$

at each time n. These coefficients are given by

$$B_\alpha(n) = \sum_{\beta=1}^{d_L} \mathbf{T}_{\alpha\beta}^{-1}(n) \left[\frac{1}{N_B} \sum_{r=1}^{N_B} (s_B^{(r)}(n) + 1) \mathbf{Y}_{A\beta}^{(r)}(n) \right],$$

$$A(n) = \sum_{\beta=1}^{d_L} B_\beta(n) \mathbf{Y}_{A\beta}^{(r)}(n) + \frac{1}{N_B} \sum_{r=1}^{N_B} s_B^{(r)}(n), \tag{6.26}$$

where

$$T_{\alpha\beta}(n) \;=\; \frac{1}{N_B}\sum_{r=1}^{N_B}[\mathbf{Y}_{A\alpha}^{(r)}(n) - \bar{\mathbf{Y}}_{A\alpha}(n)][\mathbf{Y}_{A\beta}^{(r)}(n) - \bar{\mathbf{Y}}_{A\beta}(n)],$$

$$\bar{\mathbf{Y}}_{A\alpha}(n) \;=\; \frac{1}{N_B}\sum_{r=1}^{N_B}\mathbf{Y}_{A\beta}^{(r)}(n). \qquad\qquad (6.27)$$

The collection of scalars $A(n)$ and vectors $\mathbf{B}(n)$ determine our model for the variables s_B as expressed in s_A space.

This procedure can in general be used for constructing a model for the s_A in terms of s_B or for many models of $s_B, s_C, ...,$ in terms of s_A or in terms of each other. In the exceptional case where the underlying variables are not invertible in terms of each other, this may fail, but if faced with observations alone, one can only discover this by experimenting with the procedure.

6.4.1 Lorenz Model

Breaking with our own tradition we do not display the results of predicting one variable from another in the Lorenz model. When it works, as with many things in the Lorenz system, it works very well. When it doesn't work, it is due to a special symmetry of the Lorenz system [ACP*94] and we really do not learn very much. Instead we turn immediately to our other faithful example of data from a nonlinear electrical circuit.

6.4.2 Nonlinear Circuits

Using data $V_A(n)$ and $V_B(n)$ from the hysteretic nonlinear circuit discussed in an earlier chapter, we now make models of V_B in terms of V_A and its time delays and *vice versa*. In Figure 6.7 we show the result of working in $d_E = 6$, and $d_L = 3$ and creating the local model using 10,000 data to recover V_B from V_A data. In Figure 6.8 we do just the inverse and recover V_A from V_B data. In Figure 6.9 we display the errors made in these predictions when constructing V_B as a scalar function of time delayed vectors created from V_A data. The largest error is about 10% and the average error, over the whole 1000 points predicted, relative to the RMS size of the attractor, is about 7 percent.

The ability to do this is not at all related to the systems we are discussing having chaotic motion. The idea is strictly connected to establishing a complete enough set of coordinates for the phase space of the observations at hand, then one can perform the recovery procedure just described. This means that the method may well be valuable when chaos is absent. Initially this was seen as a way to predict, and possibly control, one hard to

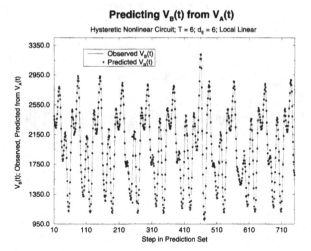

FIGURE 6.7.   The predictions of voltage $V_B(t)$ in terms of observations of $V_A(t)$ in a space of vectors formed from $V_A(t)$ and its time delays. A dimension $d_E = 6$ was used for forming the space of delay vectors from $V_A(t)$ and a time delay of $T = 6$ from average mutual information was employed. The local model was linear, and 30,000 data points were used to determine the coefficients in the local models.

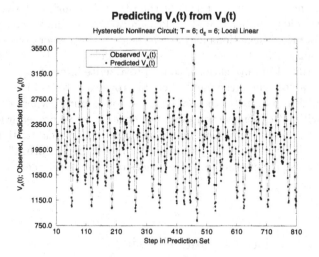

FIGURE 6.8.   The predictions of voltage $V_A(t)$ in terms of observations of $V_B(t)$ in a space of vectors formed from $V_B(t)$ and its time delays. A dimension $d_E = 6$ was used for forming the space of delay vectors from $V_B(t)$ and a time delay of $T = 6$ from average mutual information was employed. The local model was linear, and 30,000 data points were used to determine the coefficients in the local models.

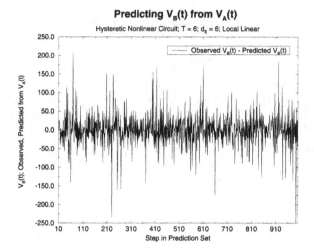

FIGURE 6.9.   The errors made in determining voltage $V_B(t)$ from measurements of $V_A(t)$ in a $d_E = 6$ dimensional space created from $V_A(t)$ and its time delays. The prediction error over the set of points where predictions were attempted was about 7%; the maximum error is about 10%.

measure variable in terms of another easy to measure variable by having a working model on one in terms of the other. Measuring the velocity in a high Reynolds number boundary layer may be possible in carefully controlled laboratory situations and not in field settings where one wishes to know the answer to that question. Making models in the laboratory, which are then utilized in applications, could prove quite interesting.

Another arena where the method could prove valuable is in the monitoring of a critical parameter in power plants, for example, where one wishes to know without expensive shutdowns or other invasive procedures when a sensitive instrument reporting that parameter is losing calibration and to continue to know the critical parameter by having other "proxy" measurements in whose reconstructed phase spaces one knows the measurement of interest.

6.5   "Black Boxes" and Physics

The measurements $s(n)$ and its few time delays provide coordinates for the effective finite dimensional space in which the system is acting after transients have died out. This gives rise to two views of modeling the dynamics from the point of view of the physicist interested in understanding the mechanisms producing the measured signals:

i The first viewpoint is that of making dynamical models in the coordinates $s(n)$ and quantities made from the $s(n)$—and doing all this in the finite dimensional space where the observed orbits evolve. This description of the system differs markedly from the original differential equations, partial or ordinary, which one might have written down from first principles or from arguments about the structure of the experiment. Verification of the model by future experiments or from examining different features of the experiment, proceeds as usual, but the framework of the models is quite different from the usual variations of Newton's laws which constitute most physics model making. The net result of this will be a nonlinear relationship or map from values of the observed variables $s(n)$ to their values $s(n+T)$ some time T later. If one of the Lyapunov exponents, as determined by the data, using methods we will describe later, is zero, then one could construct differential equations to capture the dynamics. This suggests we are confident that extrapolation to $\tau_s \to 0$, implicit in the idea of a differential equation rather than a finite time map, captures all the relevant frequencies or time scales or degrees of freedom of the source. In either case—map or differential equation—no rigid rules appear to be available to guide the modeler.

ii The second viewpoint is more traditional and consists of making models of the experiment in the usual variables of velocity, pressure, temperature, voltage, etc. The job of the scientist is to then explore the implications of the model for various invariant quantities under the dynamics such as the fractal dimensions of the attractor and the Lyapunov exponents of the system. These invariants are then the testing grounds of the models. The critical feature of the invariants is that they are not sensitive to initial conditions or small perturbations of an orbit, while individual orbits of the system are exponentially sensitive to such perturbations. The exponential sensitivity is the manifestation of the instability of all orbits in the phase space of the dynamics. Because of this intrinsic instability no individual orbit can be compared with experiment or with a computed orbit since any orbit is effectively uncorrelated with any other orbit, and numerical roundoff or experimental precision will make every orbit distinct. What remains, and what we shall emphasize in this review, is a **statistics of the nonlinear deterministic system** which produces the orbits. An additional set of invariants, *topological invariants* are also quite useful with regard to this task of characterizing physical systems and identifying which models of the system are appropriate [MSN*91].

The first modeling direction is sure to be suspect to traditional scientists whose bent is to seek mechanisms couched in terms of familiar dynamical variables. $s(t)$ may well be familiar, say a temperature or voltage, but $s(t+kT)$, even though a time lag of the familiar variable, is a nonlinear mixture of that variable and all the other dynamical quantities in the source. Models cast into differential or integral equations or discrete time maps for these quantities and the accompanying parameters in these models do provide a framework for the evolution of the system. This framework is certain to be seen as "black box" modeling by many, yet can be quite useful in providing a working description of the evolution of the system.

The second, more traditional and well tested scheme, cannot proceed without further information about the experimental setup. Fortunately this is often available. While the data set analyzed and modeled by the methods we have described can be viewed as a series of ASCII numbers which were fed into a port on one's workstation, typically we know that the data came from an experiment involving atmospheric acoustics rather than lobster stomachs, for example, and can construct relevant models in the correct context. In an example to come, involving chaotic laser intensity fluctuations, we show how this procedure can work when one exploits the additional knowledge available.

In a sense this is all to emphasize that the material discussed in this monograph does not replace scientific inquiry with a series of buttons to be clicked on with one's mouse. It provides a set of tools, and tools alone, which it is hoped are useful for the analysis of data. The results of that analysis must be interpreted using some knowledge of the source of the data, and that is where these tools will aid in scientific progress.

7
Signal Separation

7.1 General Comments

Now we address some of the problems associated with the first task in Table 1.1. **Given observations contaminated by other sources, how do we clean up the signal of interest so we can perform the analysis for Lyapunov exponents, dimensions, model building, etc.?** In linear analysis the problem concerns the extraction of sharp, narrowband linear signals from broadband "noise". This is best done in Fourier domain, but that is not the working space for nonlinear dynamics. The problem of separating nonlinear signals from one another is done in the time domain of dynamical phase space. To succeed in signal separation, we need to characterize one or all of the superposed signals in some fashion which allows us to differentiate them. This is precisely what we do in linear problems as well, though the distinction is spectral there. If the observed signal $s(n)$ is a sum of the signal we want, call it $s_1(n)$, and other signals $s_2(n), s_3(n), \ldots,$

$$s(n) = s_1(n) + s_2(n) + s_3(n) + \ldots, \qquad (7.1)$$

then we must identify some distinguishing characteristic of $s_1(n)$ which either the individual $s_i(n); i > 1$ or the sum does not possess.

The most natural of these distinguishing properties is that $s_1(n)$ or its representative in reconstructed phase space

$$\mathbf{y}_1(n) = [s_1(n), s_1(n + T_1), \ldots, s_1(n + T_1(d_{E1} - 1))] \qquad (7.2)$$

satisfies a dynamical rule: $\mathbf{y}_1(k + 1) = \mathbf{F}_1(\mathbf{y}_1(k))$ different from any dynamical rules associated with $s_2(n), s_3(n), \ldots$ If this rule is known, it can be used in ways to be discussed to estimate the sequence of $s_1(n)$ from the observations. A bound on how well one can achieve this estimate is the subject of Chapter 12.

There are three general classes of signal separation problems:

i **We know the dynamics $y_1 \rightarrow F_1(y_1)$.** [Ham90, FS89]

ii **We have observed some clean signal $y_R(k)$ from the chaotic system, and we can use the statistics of this reference signal to distinguish it from contamination.** [MA91] The reference signal may or may not be chaotic.

iii **We know nothing about a clean signal from the dynamics or about the dynamics itself.** This is the 'blind' case and requires assumptions about the signal and about the contamination for the signal separation to be effected.

It is quite helpful as we discuss work on each of these items to recall what it is we are attempting to achieve by the signal separation. Suppose the observation to be the sum of two signals: $s(n) = s_1(n) + s_2(n)$. In case (i), we want to know $s_1(n)$ with full knowledge of the dynamics $y_1 \rightarrow F_1(y_1)$. The information we use about the signal $s_1(n)$ is that it satisfies this dynamics in the reconstructed phase space, while $s_2(n)$ and its reconstruction does not. This is not really enough, since any initial condition $y_1(1)$ iterated through the dynamics satisfies the dynamics by definition. However, such a time sequence starting from an arbitrary initial condition while satisfying the deterministic rules will have nothing, in detail, to do with the desired sequence $s_1(1), s_1(2), \ldots$, which enter the observations. This is because of the intrinsic instabilities in nonlinear chaotic systems, so two different initial conditions diverge exponentially rapidly from each other as they move along the attractor. To the dynamics, then, we must add some cost function or quality function or accuracy function which tells us that we are extracting from the observations $s(n)$ that **particular** orbit which lies "closest" to the particular sequence $s_1(1), s_1(2), \ldots$, which was observed in contaminated form along with $s_2()$. If we are not interested in that particular sequence but only in properties of the dynamics, we need do nothing more since we already have the dynamics and can iterate any initial condition to find out what we want.

Similarly in case (ii), we will be interested in the particular sequence which was contaminated or masked during observation. Since we have a clean reference orbit of the system, any general or statistical question about the system is best answered by that reference orbit rather than trying to extract from contaminated data some less useful approximate orbit.

In case (iii), we are trying to learn both the dynamics and the signal "on the fly". Here it may be enough to learn the dynamics by unraveling a deterministic part of the observation in whatever embedding dimension we work in. That is, suppose the observations are composed of a chaotic signal, $s_1(k)$, which we can capture in a five dimensional embedding space combined with a two hundred dimensional contamination $s_2(k)$. If we work in five dimensions, then the second component will look non-deterministic since there will be so many false neighbors that the direction the orbit moves at nearly all points of the low dimensional phase space will look random. The distinguishing characteristic of the first signal will be its relatively high degree of determinism compared to the second.

7.2 Full Knowledge of the Dynamics

In this section we assume that the dynamics of a signal are given to us: $\mathbf{y}_1 \rightarrow \mathbf{F}_1(\mathbf{y}_1)$ in d_1 dimensional space. We observe either a contaminated d_1-dimensional signal or from a scalar measurement $s(k) = s_1(k)+s_2(k)$, we reconstruct the d_1-dimensional space. The observations are labeled $\mathbf{y}_O(k)$ in d_1 dimensions. We wish to extract from those measurements the best estimate $\mathbf{y}_E(k)$ of the particular sequence $\mathbf{y}_1(k)$ which entered the contaminated observations. The error we make in this estimation $\mathbf{y}_1(k) - \mathbf{y}_E(k) = \epsilon_A(k)$ we call the **absolute error**. It is some function of ϵ_A we wish to minimize. We know the dynamics, so we can also work with the deterministic error $\epsilon_D(k) = \mathbf{y}_E(k + 1) - \mathbf{F}_1(\mathbf{y}_E(k))$ which is the amount by which our estimate fails to satisfy the dynamics. This second error is more theoretical but since it is the deterministic evolution rule upon which we will base our signal separation methods, it is a useful quantity to monitor even though it is eventually the absolute error which we wish to make small.

A natural starting point is to ask that the square of the absolute error $|\epsilon_A(k)|^2$ be minimized subject to $\epsilon_D(k) = 0$; that is, we seek to be close to the true orbit $\mathbf{y}_1(k)$ constraining corrections to the observations by the requirement that all estimated orbits satisfy the known dynamics as accurately as possible. Using Lagrange multipliers $\mathbf{z}(k)$ this means we want to minimize

$$\frac{1}{2} \sum_{k=1}^{N} |\mathbf{y}_1(k) - \mathbf{y}_E(k)|^2 + \sum_{k=1}^{N-1} \mathbf{z}(k) \cdot [\mathbf{y}_E(k + 1) - \mathbf{F}_1(\mathbf{y}_E(k))], \qquad (7.3)$$

with respect to the $\mathbf{y}_E(m)$ and the $\mathbf{z}(m)$.

The variational problem established by this is

$$\begin{aligned}
\mathbf{y}_E(k + 1) &= \mathbf{F}_1(\mathbf{y}_E(k)), \\
\mathbf{y}_1(k) - \mathbf{y}_E(k) &= \mathbf{z}(k - 1) - \mathbf{z}(k) \cdot \mathbf{DF}_1(\mathbf{y}_E(k)),
\end{aligned} \qquad (7.4)$$

which we must solve for the Lagrange multipliers and the estimated orbit $\mathbf{y}_E(k)$. The Lagrange multipliers are not of special interest, and the critical issue is estimating the original orbit. If we wished, we could attempt to solve the entire problem by linearizing in the neighborhood of the observed orbit $y_O(k)$ and using some singular value method for diagonalizing the $N \times N$ matrix, where N is the number of observation points [FS89].

Instead we seek to recursively satisfy the dynamics by defining a sequence of estimates $y_E(k,p)$ labeled by an estimate number $p = 0, 1, 2, \ldots$ The $(p+1)^{st}$ estimate is determined by

$$y_E(k, p+1) = y_E(k, p) + \Delta(k, p), \tag{7.5}$$

where at the $p = 0$ step we choose $y_E(k, 0) = y_O(k)$, as that is all we know from observations. We will try to assure that the increments $\Delta(k, p)$ are always "small", so we can work to first order in them. If the $\Delta(k, p)$ start small and remain small, we can approximate the solution to the dynamics $x \to F_1(x)$ as

$$
\begin{aligned}
\Delta(k+1, p) + y_E(k+1, p) &= F_1(y_E(k, p) + \Delta(k, p)), \\
\Delta(k+1, p) &= F_1(y_E(k, p) + \Delta(k, p)) - y_E(k+1, p), \\
&\approx DF_1(y_E(k, p)) \cdot \Delta(k, p) + \epsilon_D(k, p), \quad (7.6)
\end{aligned}
$$

where $\epsilon_D(k, p) = F_1(y_E(k, p)) - y_E(k+1, p)$ is the deterministic error at the pth step.

We satisfy this linear mapping for $\Delta(k, p)$ by using the present value for $y_E(k, p)$. To find the first correction to the starting estimate $y_E(k, 0) = y_O(k)$ we iterate the mapping for $\Delta(k, 0)$ and then add the result to $y_E(k, 0)$ to arrive at $y_E(k, 1) = y_E(k, 0) + \Delta(k, 0)$. The formal solution for $\Delta(k, p)$ can be written down in terms of $\Delta(1, p)$. This approach suffers from numerical instability lying in the positive eigenvalues of the composition of the Jacobian matrices $DF_1(y_E(k, p))$. These are the same instabilities which lead to positive Lyapunov exponents and chaos itself. To deal with these numerical instabilities we introduce an observation due to Hammel [Ham90] and Farmer and Sidorowich [FS89]: since we know the map $x \to F_1(x)$, we can identify the linear stable and unstable invariant manifolds throughout the phase space occupied by the attractor. These linear manifolds lie along the eigendirections of the composition of Jacobian matrices entering the recursive determination of an estimated orbit $y_E(k)$. If we decompose the linear problem Equation(7.6) along the linear stable and unstable manifolds and then iterate the resulting maps **forward along the stable directions and backward along the unstable directions**, we will have a numerically stable algorithm in the sense that a small $\Delta(1, p)$ along a stable direction will remain small as we move forward in time. Similarly a small $\Delta(N, p)$ at the final point will remain small as we iterate backward in time. Combining the forward iterates and the backward iterates after this procedure will give us our next estimate of the correction to the contaminated orbit.

Each iterate of the linear map for the $\Delta(k,p)$ will fail to satisfy the condition $\Delta(k,p) = 0$ which would be the completely deterministic orbit because the movement toward $\Delta(k,p) = 0$ exponentially rapidly along both stable and unstable directions is bumped about by the deterministic error at each stage. If that deterministic error grows too large, then we can expect the linearized attempt to find the deterministic orbit closest to the observations to fail. At each iteration this deterministic error is a measure of the signal to noise level in the estimate relative to the original signal. If this level is too large, the driving of the linear system by the deterministic error will move the $\Delta(k,p)$ away from zero and keep them there. This means that the stable iteration scheme just described may well fail in a rather ungraceful fashion as the contamination level grows.

The main mode of failure of this procedure, which in its heart is a Newton-Raphson method for solving the nonlinear map, arises when the stable and unstable manifolds are nearly parallel for then one may accidentally be iterating a piece of the forward answer backward, and this is numerically unstable. In practice these directions are never precisely parallel because of numerical round-off, but if they become so close to parallel that the forward-backward procedure cannot correctly distinguish between stable and unstable directions, then there will be "leakage" from one manifold to another and this error will be exponentially rapidly magnified until the basic stability of the manifold decomposition iteration can catch up with this error.

Our examples of manifold decomposition will begin with the Hénon map of the plane to itself

$$\begin{aligned}
x(n+1) &= 1.0 + y(n) - ax(n)^2, \\
y(n+1) &= bx(n),
\end{aligned} \tag{7.7}$$

with $a = 2.8$ and $b = 0.3$. In Figure 7.1 we show the phase portrait of the Hénon attractor with no noise. Figure 7.2 shows the same data with 2.9% noise added to the data. This percentage is the RMS level of the uniformly distributed noise added to the Hénon signal relative to the RMS size of the attractor. We use the details of the Hénon map as discussed above to identify the stable and unstable manifolds everywhere along the attractor. Numerically we find the unstable manifold at $(x(j), y(j))$ by making a small perturbation at $(x(j-L), y(j-L))$ and follow it L steps forward to the point of interest. Typically we have taken $L = 15$, though as few as five or so steps will suffice. The small perturbation points with exponential accuracy in the direction of the unstable manifold at $(x(j), y(j))$. The stable manifold is found by starting L ahead of the point of interest and then iterating a perturbation to the map backward to $(x(j), y(j))$.

In Figure 7.3 we show the **absolute** error $\log_{10}[\|\mathbf{y}_E(k,p) - \mathbf{y}_1(k)\|]$ for $p = 35$ as a function of k which results from iterating this manifold decomposition procedure for 35 times. One can see the regions where the

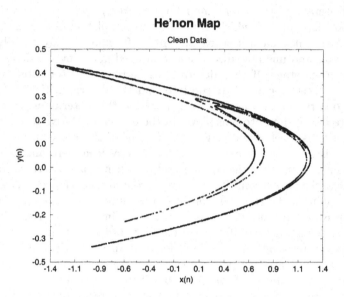

FIGURE 7.1.   Clean data from the Hénon map shown as a phase portrait in $(x(n), y(n))$ space.

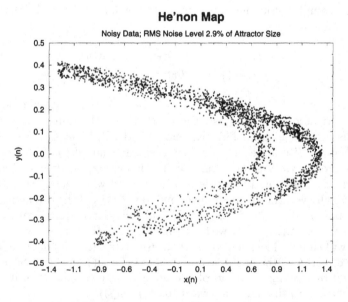

FIGURE 7.2.   Data from the Hénon map to which uniformly distributed random numbers have been added with an RMS amplitude 2.9% that of the Hènon signal.

He'non Map

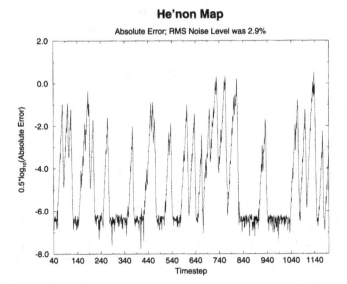

FIGURE 7.3.   The absolute error between the true input data from the Hénon map and the result estimated from using manifold decomposition methods for the separation of the Hénon signal from the random numbers as shown in Figure 7.2. When the stable and unstable manifolds in the map come close to being parallel, the method loses accuracy and "glitches" in the estimate occur. The method recovers and accurately estimates the original clean signal.

stable and unstable manifolds become almost parallel, called *homoclinic tangencies*, and then see the orbit recover. In Figure 7.4 we show the phase portrait of the cleaned up data. Much of the structure of the Hénon attractor has been recovered, though we can see some outliers associated with noise levels which bounce the orbit far from the attractor and from which the manifold decomposition method never recovers. If the noise level were reduced to 1% or so, we would not see these outliers, and the absolute error would always be much, much less. It is useful to see the sensitivity of the method to even relatively small levels of signal contamination.

The "glitches" caused by the homoclinic tangencies can be cured in at least two ways. First, one can go back to the full least-squares problem stated above and in a region around the tangency do a full diagonalization of the matrices that enter. One can know where these tangencies will occur because the Jacobians, which we know analytically, since we know $\mathbf{F}_1(\mathbf{x})$, are evaluated at the present best estimate $\mathbf{y}_E(k,p)$. An alternate solution is to scale the step $\Delta(k,p)$ we take in updating the estimated trajectory by a small factor in all regions where a homoclinic tangency might occur. One can do this automatically by putting in a scale factor proportional to the angle between stable and unstable directions or by doing it uniformly

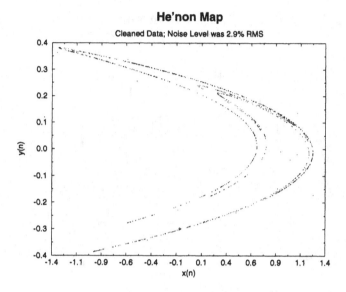

FIGURE 7.4.   The phase portrait in $(x(n), y(n))$ space for the cleaned up signal from the Hénon map as seen in Figure 7.2. There are outliers, but one sees the shape and structure of the attractor recovered from the noisy data.

across the orbit. The penalty one pays is a larger number of iterations to achieve the same absolute error, but the glitches are avoided.

In any of these cases one arrives at an estimate of the true orbit $\mathbf{y}_1(k)$ which is accurate to the round-off error of one's machine. So if one has an orbit of a nonlinear dynamical system which in itself is of some importance but it is observed contaminated by another signal, one can extremely accurately extract that orbit if the dynamics are known. The limitations on the method come from the relative amplitudes of the contamination and the chaotic signal. In practice when the contamination rises to 10% or so of the chaos, the method works unreliably. The other important application of the method would be using the chaotic signal as a mask for some signal of interest which has no relation to the dynamics itself. Then one is able to estimate the chaotic signal contained in the observations extremely accurately and by subtraction expose the signal of interest. Since this application would be most attractive when the amplitude of the signal of interest is much less than the masking chaos, the method should perform quite well. In Figure 7.5 we show a signal composed of a sequence of square pulses which was buried in a signal from the Hénon map with an original signal to chaos ratio of about 4500 to 1 (or -73.1 dB in power). The signal was extracted from the chaos with the resulting waveform. The signal to residual noise ratio after extraction from the Hénon chaos was $+16.5$ dB. The power of this method to extract small signals implanted in known

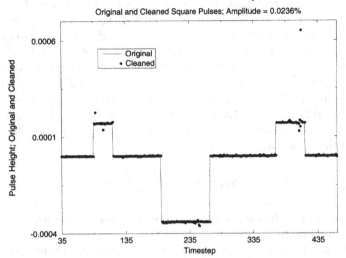

FIGURE 7.5.   The original and cleaned sequence of square pulses contaminated by Hénon chaos and cleaned by the manifold decomposition method. The height of the pulses is 0.0236% of the RMS amplitude of the masking Hénon signal. In the manifold decomposition method the smaller the signal of interest, that is the smaller the deviation from the known chaotic signal, the better cleaning can recover the signal of interest.

chaos is quite good. Indeed, to the extent the signal strength is above the round-off error of the computer, the smaller the better. Indeed since the signal of interest, here a few square pulses for illustration is the noise added to the known Hénon chaos, if it becomes too large relative to the chaotic masking signal, the manifold decomposition method will fail. So this method is only good for removing masking from signals which are quite small compared to the masking chaos. Further, implementing this method with chaos in more than two dimensions can be quite complicated as one must rather accurately determine the stable and unstable manifolds throughout phase space, and one must determine all of them in higher dimensions. Nonetheless, the implications for private communications are quite clear.

7.3   Knowing a Signal: Probabilistic Cleaning

The second method of signal separation [MA91] rests on having observed a clean signal from a system at some earlier time. This reference signal $y_R(k)$ is recorded and used to establish statistical properties of the motion along the attractor. Later one receives another signal $\mathbf{y}_C(k)$ from the same system along with some contamination $\mathbf{z}(k)$. The observed signal is $\mathbf{s}(k) = \mathbf{y}_C(k) + \mathbf{z}(k)$. We have no *a priori* knowledge of the contamination $\mathbf{z}(k)$ except that it is independent of the clean signal $\mathbf{y}_C(k)$. Except for coming from the same dynamics the two signals $\mathbf{y}_R$ and $\mathbf{y}_C$ are totally uncorrelated because of the chaos itself. The probabilistic cleaning procedure seeks to estimate that clean sequence $\mathbf{y}_C(n)$ which maximizes the conditional probability $P(\{\mathbf{y}_C\}|\{\mathbf{s}\})$ that having observed the contaminated sequence $\mathbf{s}(1), \mathbf{s}(2), \ldots, \mathbf{s}(N)$ the real sequence was $\mathbf{y}_C(1), \mathbf{y}_C(2), \ldots, \mathbf{y}_C(N)$. To determine the maximum over the estimated values of the $\mathbf{y}_C$ we write this conditional probability in the form

$$P(\{\mathbf{y}_C\}|\{\mathbf{s}\}) = \frac{P(\{\mathbf{s}\}|\{\mathbf{y}_C\})P(\{\mathbf{y}_C\})}{P(\{\mathbf{s}\})}, \tag{7.8}$$

and note that only the entries in the numerator enter in the maximization over estimated clean orbits.

The probability of the sequence of clean observations can be estimated using the fact that $\mathbf{y}_C(k+1) = \mathbf{F}(\mathbf{y}_C(k))$ so it is first order Markov in the correct multivariate space where the local dynamics acts. This allows us to write

$$P(\mathbf{y}_C(1), \mathbf{y}_C(2), \ldots, \mathbf{y}_C(m)) =$$
$$P(\mathbf{y}_C(m)|\mathbf{y}_C(m-1))P(\mathbf{y}_C(1), \mathbf{y}_C(2), \ldots, \mathbf{y}_C(m-1)) \tag{7.9}$$

and use kernel density estimates [Sil86] of the Markov transition probability $P(\mathbf{y}_C(m)|\mathbf{y}_C(m-1))$ via

$$P(\mathbf{y}_C(m)|\mathbf{y}_C(m-1)) \propto \frac{\sum_{j',j=1}^{N} K(\mathbf{y}_C(m) - \mathbf{y}_R(j))K(\mathbf{y}_C(m-1) - \mathbf{y}_R(j'))}{\sum_{n=1}^{N} K(\mathbf{y}_C(m-1) - \mathbf{y}_R(n))}. \tag{7.10}$$

The conditional probability $P(\{\mathbf{s}\}|\{\mathbf{y}_C\})$ is estimated by assuming that the contamination $\mathbf{z}$ is independent of the $\mathbf{y}_C$, so

$$P(\mathbf{s}(j); j = 1, m|\mathbf{y}_C(k); k = 1, m) = \prod_{k=1}^{m-1} P_z(\mathbf{y}_C(k) - \mathbf{s}(k))$$
$$= P(\mathbf{s}(j); j = 1, m-1|\mathbf{y}_C(k); k = 1, m-1)P_z(\mathbf{y}_C(m) - \mathbf{s}(m)) \tag{7.11}$$

with $P_z(\mathbf{x})$ the probability distribution of the contaminants.

These forms for the probabilities lead to a recursive procedure for searching phase space in the vicinity of the observations $\mathbf{s}$. One searches for adjustments to the observations $\mathbf{s}$ which maximize the conditional probability $P(\{\mathbf{y}_C\}|\{\mathbf{s}\})$. The procedure is performed recursively along the observed orbit with adjustments made at the end of each pass through the sequence. Then the observations are replaced by the estimated orbit, the size of search space is shrunk down toward the orbit by a constant scale factor, and the

Lorenz Attractor

r = 45.92; σ = 16.0; b = 4.0

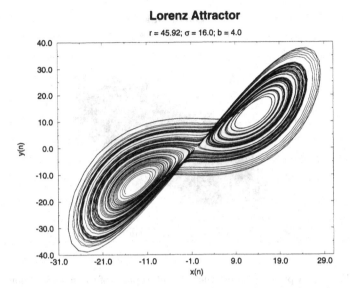

FIGURE 7.6.   A projection onto the $(x(t), y(t))$ plane of the Lorenz attractor
for a clean signal. The parameters of the Lorenz system, Equation (3.5), were $r =
45.92, b = 4.0$, and $\sigma = 16.0$.

procedure is repeated. This search procedure is similar to a standard *maximum a posteriori* [Tre68] technique with two alterations: (1) the probabilities required in the search are not assumed beforehand, they are estimated
from knowledge of the reference orbit $\{y_R(k)\}$, and (2) in the search procedure one systematically shrinks the initial volume in state space in which
a search is made.

The nonlinear dynamics enters through the reference orbit and the use
of statistical quantities of a deterministic system. The method is cruder
than the manifold decomposition algorithm, but it also requires much less
knowledge of the dynamical system. It works when the signal to noise ratio
is as low as 0 dB for extracting an initial chaotic signal contaminated by
noise which is identically and independently distributed. For a signal of
interest buried in chaos, it has proven effective when the signal to chaos
ratio is as low as −20 dB, and even smaller signals will work. In Figure 7.6
we show a sample of the Lorenz attractor projected onto the $(x(t), y(t))$
plane before contamination. In Figure 7.7 we have the contaminated signal
with uniform random numbers distributed in the range [−12.57, 12.57]
added to each component. The signal (Lorenz chaos) to noise ratio is
0.6 dB. We show in Figure 7.8 the signal recovered by the probabilistic
cleaning method after 30 passes through the algorithm. In implementing
this cleaning of the quite contaminated Lorenz signal, we used 5500 points
on a clean reference orbit and cleaned about 2000 points shrinking the

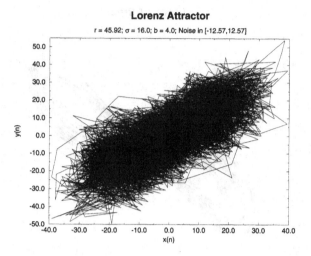

FIGURE 7.7.   A projection onto the $(x(t), y(t))$ plane of the same signal as in Figure 7.6 when uniformly distributed random numbers in $[-12.57, 12.57]$ were added to the clean Lorenz signal. The signal to noise ratio with this level of contamination is about 0.6 dB.

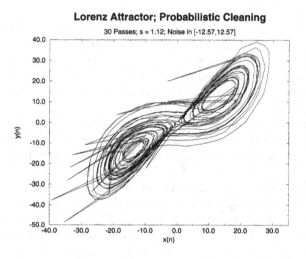

FIGURE 7.8.   A projection onto the $(x(t), y(t))$ plane of the Lorenz attractor recovered from the data seen in Figure 7.7 using 25,000 points on a reference orbit, 30 passes of the scaled probabilistic cleaning method with a 12% shrinkage of searched phase space at each pass. There are outliers, but the structure of the attractor is well recovered. This contamination level is far beyond the ability of the manifold decomposition method to deal with.

Pulse In Chaotic Lorenz Signal

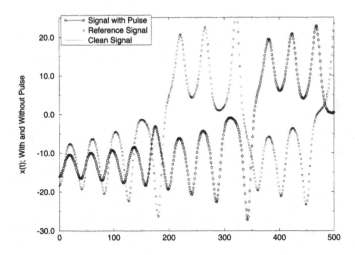

FIGURE 7.9.   A pulse of amplitude 4.71 in the neighborhood of time step 175 has been added to the $x(t)$ dynamical variable of the Lorenz model, Equation (3.5). This figure shows the original signal (solid line), the signal with the added pulse (circles joined by a solid line), and the reference signal, from the Lorenz system (in circles alone) used in the scaled probabilistic cleaning procedure.

region of phase space searched at each step by a factor of 12%. Clearly there are outliers which can often be captured by shrinking the phase space volumes at each pass through the algorithm more slowly. This method can deal with much higher noise levels than manifold decomposition, requires much less knowledge of the dynamics of the clean signal, and works well, though slowly, in higher dimensions. It is cruder than manifold decomposition, but uses less knowledge as well. In Figure 7.9 we show a pulse added to the Lorenz attractor with an initial pulse to chaos ratio of -8.7 dB. Also shown in this figure is the behavior of the original Lorenz signal before the pulse was added and the reference orbit used in the probabilistic cleaning process. The reference orbit is completely uncorrelated with the original orbit as it should be for any two distinct samples from the attractor. Figure 7.10 is a blowup of the section of these signals where the pulse was added, near time step 175. The pulse was extracted from the Lorenz chaos to give the result shown in Figure 7.11 which has a signal to noise ratio of 54.8 dB relative to the known initial pulse. This method can clearly extract small signals out of chaos when a good reference orbit is available. If there were information contained in this spike and it had been hidden by chaos of which we had a sample from the past—the reference orbit—we would have recovered the masked information by this probabilistic cleaning method.

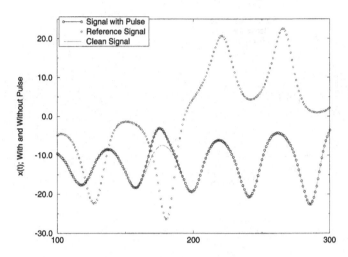

FIGURE 7.10.   An enlargement of Figure 7.9 so the effect of the pulse near time step 175 can be seen more clearly.

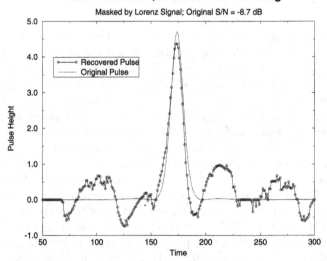

FIGURE 7.11.   The recovered and original pulse added to the $x(t)$ dynamical variable of the Lorenz system. The pulse was recovered using the scaled probabilistic cleaning method. The original signal to noise ratio was -8.7 dB. After cleaning, the signal to noise ratio is about $+10$ dB.

Probabilistic cleaning and manifold decomposition both use features of the chaos which are simply not seen in a traditional Fourier treatment of signal separation problems. In conventional approaches to noise reduction the noise is considered unstructured stuff in state space with some presumed Fourier spectrum. Here we have seen the power of knowing that chaotic signals possess structure in state space even though their Fourier spectra are continuous, broadband and appear as noise to the linear observer. We are confident that this general observation will prove the basis for numerous applications of nonlinear dynamics to signal based problems in data analysis and signal synthesis or communications.

7.4 "Blind" Signal Separation

When we have no foreknowledge of the signal, the dynamics, or the contaminant, we must proceed with a number of assumptions explicit or implicit. We cannot expect any method to be perfectly general, of course, but within some broad domain of problems we can expect to succeed in separating signals with some success.

In the work which has been done in this area there have been two general strategies:

i **make local polynomial maps** using neighborhood to neighborhood information. Then adjust each point to conform to the map determined by a collection of points; i.e., the whole neighborhood. In some broad sense a local map is determined with some kind of averaging over domains of state space, and after that this averaged dynamics is used to realign individual points to a better deterministic map. The work of Kostelich and Yorke [KY90] is of this kind.

ii **use linear filters locally or globally** and then declare the filtered data to be a better "clean" orbit. Only moving average or FIR filters will be allowed, or we will have altered the state space structure of the dynamics itself. The work of Landa and Rosenblum [LR89], Cawley and Hsu [CH92], Sauer [Sau92], and Pikovsky [Pik86] all fall in this class.

In the first approach one makes a model of the dynamics of the signal by representing it as a deterministic rule in phase space. The rule is implemented as a local neighborhood to neighborhood map and the residual in this operation is associated with the contamination to be removed. In the second approach, some attribute is associated with the contamination, and then this is filtered out in some way or another.

The first of these blind cleaning methods proceeds in the following fashion. Select neighborhoods around every data point $\mathbf{y}(n)$ consisting of N_B

points $\mathbf{y}^{(r)}(n); r = 1, 2, \ldots, N_B$. Using the $\mathbf{y}^{(r)}(n)$ and the $\mathbf{y}(r; n + 1)$ into which they map, form a local polynomial map

$$
\begin{aligned}
\mathbf{y}(r; n + 1) &= \mathbf{g}(\mathbf{y}^{(r)}(n), \mathbf{a}), \\
&= \mathbf{A} + \mathbf{B} \cdot \mathbf{y}^{(r)}(n) + \mathbf{C} \cdot \mathbf{y}^{(r)}(n)\mathbf{y}^{(r)}(n) + \cdots, \quad (7.12)
\end{aligned}
$$

where the coefficients are determined by a local least squares minimization of the residuals in the local map. Now focusing on successive points along the orbit $\mathbf{y}(n)$ and $\mathbf{y}(n + 1)$ find small adjustments $\delta\mathbf{y}(n)$ and $\delta\mathbf{y}(n + 1)$ which minimize

$$
\|\mathbf{y}(n + 1) + \delta\mathbf{y}(n + 1) - \mathbf{g}(\mathbf{y}(n) + \delta\mathbf{y}(n))\|^2, \quad (7.13)
$$

at each point on the orbit. The adjusted points $\mathbf{y}(k) + \delta\mathbf{y}(k)$ are taken as the cleaned orbit.

In the published work one finds the use of local linear maps, but the generalization seems clear. Indeed, it seems to us that use of local measure based orthogonal polynomials $\phi_a(\mathbf{x})$ would be precisely in order here as a way of capturing, in a relatively contamination robust fashion, a good representation of the local dynamics which would then be used to adjust orbit points step by step. Using only local linear maps is in effect a local linear filter of the data which does not use any special features of the dynamics, such as its manifold structure or its statistics, to separate the signal from the contamination. We suspect that the order of signal separation is limited to $O(\frac{1}{\sqrt{N_B}})$ as a kind of central limit result. The arguments in Section 12.3 would seem to support this as well. In that section we evaluate a general kind of bound on how well one might estimate a signal in the presence of noise when not using specific dynamical properties of the signal. This is actually quite sufficient for some purposes, though is unlikely to allow for separation of signals which have significant overlap in their Fourier spectrum or for separation of signals to recover a useful version of the original, uncontaminated signal as one would desire for a communications application.

The second approach to separating signals using no special knowledge of the dynamics rests on the intuition that a separation in the singular values of the local or global sample covariance matrix of the data will occur between the signal, which is presumed to dominate the numerically larger singular values and the noise which is presumed to dominate the smaller singular values. The reasoning is more or less that suggested in an earlier section: the noise contributes equally to all singular values since the singular values of the sample covariance matrix measures how much data, in a least squares sense, lies in the eigendirection associated with that singular value. If the data is embedded in a rather large dimension $> d_E$, the singular values with the largest index are populated only by the noise and the lower index singular values are governed by the data plus some contamination. While this may be true of white noise, it is quite unlikely for 'red' noise which has most of its power a low frequencies. Nonetheless given this limitation, the method may work well.

The idea is to form the sample covariance matrix in dimension $d > d_E$ and then project the data (locally or globally) onto the first d_S singular values. This is then taken as new data and a new time delay embedding is made. The procedure is continued until the final version of the data is "clean enough". Landa and Rosenblum [LR89] project onto the direction corresponding to the largest singular value and argue that the amount of 'noise reduction' is proportional to $\frac{d_E}{d}$ multiplied together the number of times the procedure is repeated. Cawley and Hsu [CH92] do much the same thing, with interesting twists on how the "new" scalar data is achieved, but quite importantly the projection is done locally in phase space. The projection onto the singular directions is a local linear filter of the data. A combination of local filters, each different, makes for a global nonlinear filter, and in that sense the method of Cawley and Hsu represents a step forward from the earlier work of Landa and Rosenblum. Pikovsky [Pik86] also does a form of local averaging, but does not use the local singular value structure in it.

All in all one can say that there is much exploration yet to be done on the properties and limitations of blind signal separation. The hope is certainly that a method, perhaps based on invariant measure orthogonal polynomials, that utilizes the local nonlinear structure of the data will have some general applicability and validity.

7.5   A Few Remarks About Signal Separation

The very interesting theoretical work on signal separation has not been enormously successful in its application to its original goal: reduction of noise in observations of chaotic systems. It is clear that were one to know much about the uncontaminated signal, its actual dynamics or an earlier reference orbit from which its clean statistics could be determined, one can do rather well in the signal separation. This is likely to be useful when one really wishes to know, with some precision, about the signal which was contaminated. To this end one can see a role for the first of these separation methods in communications. Suppose one knows a reference orbit from earlier sampling of some chaotic data or knows in detail the dynamics of a chaotic signal generator. To an output signal from another source $s(t)$ we add a signal from the chaotic generator $c(t)$ and we give ourselves the task of reading the signal $s(t)$ from the received sum $s(t) + c(t)$. Using either of the first methods would allow us to solve this problem, though perhaps not in real time as the computations required can be quite lengthy. The method would be to estimate the chaotic signal with $c_E(t)$ and then by subtraction allow an estimate of the original signal $s(t) + c(t) - c_E(t)$.

A potential application of this could be in cleaning up signals propagated through the ocean or atmosphere when those signals are contaminated by those communication channels. If we can characterize the channels by their action on the signal as resulting in additive contamination to the signal, then previous measurements of that contamination would allow recovery of the clean signal. If we know the dynamical rules for a communications channel, that is the actual equations of motion, then the manifold decomposition method applied to the linear stable and unstable manifolds of that dynamics would equally well allow the recovery of clean signals.

These potential applications turn the original problem on its head. Instead of recovering a chaotic signal of special interest by removing unwanted noise which could be quite high dimensional, we seek situations where another signal of interest, perhaps low dimensional, is contaminated by a known or slightly known (only a reference orbit is known) chaotic communications channel. These applications seem rarefied but often interesting.

As to the recovery of a chaotic signal when it is contaminated by high dimensional 'noise', one has to make some rather dramatic assumptions about the nature of the noise. When these are borne out, the methods can work rather well. However, if one has assumed that the noise is white, Gaussian and it turns out to be red and Poisson distributed and have long tails in its correlation function, one is unlikely to do as well. It may be that locally filtering the contaminated signal, using some sort of low pass filter, will do just as well as more dynamically based approaches, such as manifold decomposition, which substantial foreknowledge. After all if one has a chaotic signal from some source and one wishes to classify that source using fractal dimensions or Lyapunov exponents, one doesn't really require the precise clean signal that happened to become polluted by noise or other chaos. Any signal would do, including the nice clean reference signal or any signal from known dynamics, if one knows that.

8
Control and Chaos

Control and chaos in the same phrase would seem contradictory, but the reader knows by now that chaos is both predictable and structured in phase space. That phase space structure contains within it many simpler topological features which are unstable [DN79] and through which the system may have passed as we altered the driving on the system by varying parameters. It may not come as a surprise then that careful adjustment of these same parameters using the known stable and unstable directions of the vector field might allow one to stabilize what had become unstable. In this chapter we examine a very interesting set of methods for control which slightly change the vector field as given, by adding dynamical degrees of freedom to the system. These make formerly constant parameters time dependent. After doing this phase space structures "nearby" the unstable structures of the original system can now become stable structures of the augmented system. In particular unstable periodic orbits of the original system may be slightly altered but acquire stability as the vector field develops a carefully selected set of time dependencies. Another kind of control scheme we will examine traces its lineage to standard control theory [Owe81, Bro85] even though that may often cast in a linear context. In this approach an external force acting on the dynamical variables is added to the vector field while the parameters of the original dynamics are unaltered.

In either case one alters the vector field allowing motions previously unstable to become stabilized or allowing motions desired by the user to become possible. Both methods can also be seen as synchronization of the original dynamics by external forces designed to place the combined system on a submanifold where a selected form of regular behavior is stable. These forces act either on the parameters or the dynamical variables in the vector field of the original system, enlarge the number of degrees of freedom in the dynamics, and change the evolution of orbits.

The main discussion of control in nonlinear systems (it really has little to do with chaos) deviates from the central theme of this book which is analysis of data from chaotic systems except when we seek to apply the methods to observed systems whose detailed dynamical equations are unknown to us. In that instance, which is essentially always, we need the tools discussed earlier to capture the correct state space in which to apply the control methods.

8.1   Parametric Control to Unstable Periodic Orbits

First we address control by altering system parameters. The usual goal is moving a chaotic system from irregular to periodic, regular behavior. The method has two elements

   i the first is an application of relatively standard control theory methods beginning with a linearization around the orbits, and

   ii the second is an innovative idea on how to utilize properties of the **nonlinear** dynamics to achieve the linearized control in an exponentially rapid fashion. One of the keys to the method is detailed knowledge of the phase space and dynamics of the system. This can be gleaned from observations on the system in the manner we have outlined throughout this monograph. An attractive feature of the method is that it uses small changes in parameter values to achieve the control once the system has been maneuvered into the appropriate linearized regime by the nonlinear aspects of the problem. The necessity for only small parameter changes comes from the use of the stable manifold of the regular orbits to which one chooses to move the system. Along this manifold the desired regular motion is approached exponentially rapidly. If one changed the parameters by large amounts, the phase space structure of the unaltered system might be substantially changed as well, and the method relies on already knowing that structure from observations of the chaotic dynamics. So small parameter changes are quite attractive. Further, in any practical sense, since work is done in changing parameters, restricting oneself to small changes means small forces acting on the chaotic system. The intrinsic instabilities of the system do the work for you.

We illustrate the parametric control method using discrete time dynamics: $\mathbf{v}(n) \rightarrow \mathbf{v}(n+1) = \mathbf{F}(\mathbf{v}(n), p)$ where p is a parameter which we can adjust by external forces. Within a hyperbolic strange attractor is a dense set of unstable periodic orbits [DN79]. A "hyperbolic" strange attractor is an idealization of realistic, physically realizable chaotic motions. Hyperbolicity assumes that every point in phase space has nonparallel stable and unstable local linear manifolds. This is to say that no homoclinic tangencies are present. No example from physical systems has this property, but when the vector field gives rise to chaotic motions, this property is satisfied essentially everywhere in the phase space. The presence of these unstable periodic orbits is a characteristic sign of chaotic motion. The goal of the control algorithm is to move the system from chaotic behavior to motion along a selected periodic orbit. The orbit is **unstable** in the original, uncontrolled system, but it is **stable** in the controlled system. The controlled system has a larger phase space since the parameter which was fixed in the original dynamics is now time dependent and thus its role alters from parametric to dynamical. So one or more "extra" equations are required to specify the time variation of the parameter $p(n)$. In the new $(\mathbf{v}(n), p(n))$ dynamics the unstable periodic orbit with fixed $p(n)$ becomes stable.

The first step in solving this problem is finding the locations of the unstable periodic orbits [ACE*87, LK89]. If we are given an orbit of a dynamical system $\mathbf{y}(n)$, we can locate the unstable periodic orbits within it by asking for near recurrences of orbits. For this we ask when, if at all, an orbit returns to a close neighborhood of its value at some time n. "Close" is defined using a small, positive number ϵ: for each data vector $\mathbf{y}(n)$ in the original dynamics find the **smallest** time index $k > n$ such that

$$\|\mathbf{y}(n) - \mathbf{y}(k)\| < \epsilon,$$

and using Euclidian distance in the $\mathbf{y}(n)$ space is fine. This time $k - n$ is the recurrence time of the orbit. Since this orbit and every orbit is unstable, we cannot realistically expect to see recurrence with finite data sets, if ϵ is too small, and we cannot expect the orbit to keep recurring in a small ϵ ball very many times as the instability will drive it off to another part of phase space. However, we can expect the orbit to populate a reasonable sized ϵ ball a few times before departing to parts unknown. In practice choosing ϵ a few percent of the overall size of the attractor will work [LK89].

The period of this orbit is $M = k - n$. We expect other orbits with periods which can be multiples of M. An example of a recurrence plot which identifies unstable periodic orbits from chaotic data is plotted in Figure 8.1 and shows the number of points with recurrence time P against times P using experimental data collected from the Belousov-Zhabotinski chemical reaction [RSS83, CMN*87]. Spikes in this plot indicate that there is a fundamental periodic orbit $\mathbf{y}_P(1) \mapsto \mathbf{y}_P(2) \mapsto, \cdots, \mapsto \mathbf{y}_P(M) = \mathbf{y}_P(1)$ near $M = 125$. The discrete time vectors $\mathbf{v}(n)$ are $\mathbf{y}_P(nM)$ for $n = 1, 2, \ldots$ for orbits with periods which are multiples of the minimum M.

When the system is driven with a periodic external force, finding unstable periodic orbits associated with the period of the driving force is straightforward. We again look for recurrent points, but now in the surface of section

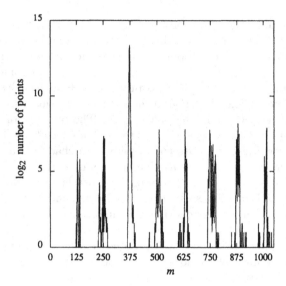

FIGURE 8.1.   A recurrence plot for data from the Belousov-Zhabotinski chemical reaction [LK89]. The plot displays the number of times points in phase space recur against the time of recurrence. Spikes in this plot indicate the presence of unstable periodic orbits in the attractor.

chosen at the period of the forcing. Typically for driven systems with data acquired once every period of the drive we will find $M = 1$. There is only one major difference between this set of unstable periodic orbits and those collected using ϵ balls in phase space. In the previous method the data vectors between $\mathbf{y}_P(1)$ and $\mathbf{y}_P(M) = \mathbf{y}_P(1)$ give the trajectory covered by the periodic orbit. For data that is taken only once every period of the drive we usually don't have the details of the trajectory with the lowest period as we do not sample those points.

The dynamics is $\mathbf{v}(n+1) = \mathbf{F}(\mathbf{v}(n), p)$, and the exact phase space location in $\mathbf{v}$ of the periodic orbit $\mathbf{v}_P \mapsto \mathbf{v}_P$ is a function of p. For the selected unstable periodic orbit, we call $p = p_0$. When the orbit of the original system comes "close" to the unstable periodic orbit of interest $\mathbf{v}_P(p_0)$ we want to alter p in the neighborhood of p_0 to stabilize this. Since we want to restrict ourselves to small changes in p and small distances from the unstable periodic orbit we have selected to stabilize, one must either wait until the experimental chaotic orbit comes near this or use some dynamical technique to force it into this region [SOGY90, SDG*92]. Once $\mathbf{v}(n)$ comes near $\mathbf{v}_P(p_0)$ the control is given by the following steps [OGY90] which we describe in two dimensions though the method is generalizable.

We have a new parameter $p(n) = p_0 + \delta p(n)$ with the time dependence on the parameter indicating we can change p at each time step, if we wish. We want to linearize the map in the neighborhood of the unstable periodic orbit and the parameter p_0. We can expand the map $\mathbf{v}(n + 1) = \mathbf{F}(\mathbf{v}(n), p(n))$

in the neighborhood of any of the members of the unstable periodic orbit. Call the point around which we expand $\mathbf{v}_0$. When we alter p, the location of this point goes to $\mathbf{v}_0(p) = \mathbf{v}_0 + \mathbf{g}\delta p$, where the vector $\mathbf{g}$ is

$$\mathbf{g} = \frac{\partial \mathbf{v}_0(p)}{\partial p}, \tag{8.1}$$

evaluated at p_0. $\mathbf{g}$ is determined by experimentally or computationally altering the parameter p in the neighborhood of p_0 and explicitly observing the movement of the point $\mathbf{v}_0$.

By regarding the map $\mathbf{F}(\bullet)$ as the Mth iterate of some basic map $\mathbf{v} \to \mathbf{f}(\mathbf{v})$ so that the point $\mathbf{v}_0$ on the periodic orbit of period M becomes a fixed point of the new map $\mathbf{v}_0 = \mathbf{F}(\mathbf{v}_0)$, we can simplify our algebra. Every point of the unstable periodic orbit of period M is a fixed point of this iterated map. Now we want to work in the neighborhood of $\mathbf{v}_0$, so we expand the iterated map in two dimensions $\mathbf{v}(n+1) = \mathbf{F}(\mathbf{v}(n), p(n))$ around both $\mathbf{v}_0$ and p_0. This leads to

$$
\begin{aligned}
\mathbf{v}(n+1) &= \mathbf{v}_0 + \mathbf{w}(n+1) \\
&= \mathbf{F}(\mathbf{v}(n)) \\
&= \mathbf{F}(\mathbf{v}_0 + \mathbf{w}(n)) \\
&= \mathbf{v}_0 + \mathbf{DF}(\mathbf{v}_0) \cdot \mathbf{w}(n) + \frac{\partial \mathbf{F}}{\partial p} \delta p(n), \tag{8.2}
\end{aligned}
$$

where $\mathbf{w}(n)$ is a small phase space displacement. Using $\mathbf{v}_0(p) = \mathbf{F}(\mathbf{v}_0(p), p)$, so

$$\frac{\partial \mathbf{F}}{\partial p} = \mathbf{g} - \mathbf{DF}(\mathbf{v}_0) \cdot \mathbf{g}, \tag{8.3}$$

we arrive at

$$\mathbf{w}(n+1) = \mathbf{g}\delta p(n) + \mathbf{DF}(\mathbf{v}_0) \cdot [\mathbf{w}(n) - \mathbf{g}\delta p(n)]. \tag{8.4}$$

In two dimensions we have one unstable direction $\hat{\mathbf{u}}$ and one stable direction $\hat{\mathbf{s}}$. These are right eigendirections of the Jacobian matrix $\mathbf{DF}(\mathbf{v}_0)$. This matrix also has left unstable and stable directions $\mathbf{f_u}$ and $\mathbf{f_s}$. As usual $\mathbf{f_u} \cdot \hat{\mathbf{u}} = 1$, $\mathbf{f_u} \cdot \hat{\mathbf{s}} = 0$, $\mathbf{f_s} \cdot \hat{\mathbf{s}} = 1$, and $\mathbf{f_s} \cdot \hat{\mathbf{u}} = 0$, and the spectral decomposition of $\mathbf{DF}(\mathbf{v}_0)$ is

$$\mathbf{DF}(\mathbf{v}_0) = \lambda_u \hat{\mathbf{u}} \mathbf{f_u} + \lambda_s \hat{\mathbf{s}} \mathbf{f_s}, \tag{8.5}$$

where the stable eigenvalue λ_s lies within the unit circle, and the unstable eigenvalue λ_u is outside the unit circle.

The control scheme is to choose $\delta p(n)$ so $\mathbf{w}(n+1)$, the deviation from the point $\mathbf{v}_0$ at the next iteration of the map, is along $\hat{\mathbf{s}}$. Thus we want $\mathbf{f_u} \cdot \mathbf{w}(n+1) = 0$, and this translates to

$$\delta p(n) = \frac{\lambda_u}{\lambda_u - 1} \left[\frac{\mathbf{w}(n) \cdot \mathbf{f_u}}{\mathbf{g} \cdot \mathbf{f_u}} \right]. \tag{8.6}$$

This is a straightforward rule to implement once one has done the required work to establish the vector **g** and the eigenvalues and eigenvectors of the matrix $\mathbf{DF}(\mathbf{v}_0)$. One knows the location of the unstable periodic orbit points $\mathbf{v}_0$, and when the orbit comes close to that point at time n, namely, $\mathbf{w}(n)$ is small, we shift the parameter from p_0 to $p_0 + \delta p(n)$. This, as just described, is designed to drive the vector $\mathbf{w}(n+1)$ onto the stable manifold of $\mathbf{v}_0$, and the dynamics, with p now shifted back to p_0, then tries to drive the orbit into $\mathbf{v}_0$ which is where we wish it to go. As noted, it would simply go into $\mathbf{v}_0$ exponentially rapidly and stay there except that the point $\mathbf{v}_0$ is unstable and there are always errors in the mapping or round-off errors in numerics which bump the orbit off the stable manifold and drive it away from $\mathbf{v}_0$. This means one must monitor the orbit and as it moves away from $\hat{\mathbf{u}}$, apply the correction δp again as needed. An application of this method, once the orbit comes close to the desired $\mathbf{v}_0$, involves repeated applications of the rule for $\delta p(n)$ to make sure the instabilities are overcome whenever required.

This discussion was set in known coordinates for the map. If one has constructed the dynamical space using time delay reconstruction, then the dependence of the map on time dependent parameters is no longer solely on the previous value of the parameter, but it may depend, at step $n + 1$, on $p(n)$ as well as $p(n-1)$ [ND92, RGOD92]. This simple extension of the basic result is true if the time between Poincaré sections is $\Delta t > (d_E - 1)T$. If the time between sections is less, then dependence on earlier values of $p(n)$ will enter [ND92]. These additional dependencies on earlier values of the time dependent parameter do not change the general outlook for this control by parameter variation, but they do necessitate small alterations in the rule for choosing $\delta p(n)$ to drive the map onto the stable manifold.

The control scheme can be cast into language reminiscent of standard pole placement or pole allocation methods [Owe81, Bro85]. We use the argument from Alsing, *et al.* [AGK94] to illustrate this. The relation to pole placement has been treated by Romeiras, et al. as well [RGOD92]. In the expression for the small deviation $\mathbf{w}(n + 1)$ from the fixed point $\mathbf{v}_o(p)$ given above, write

$$\mathbf{w}(n + 1) = \mathbf{DF}(\mathbf{v}_0)\mathbf{w}(n) + \mathbf{P}\delta p(n), \qquad (8.7)$$

where **P** is $\frac{\partial \mathbf{F}}{\partial p}$. Now choose $\delta p(n)$ proportional to the deviation $\mathbf{w}(n)$ from the fixed point

$$\delta p(n) = -\mathbf{K}\mathbf{w}(n), \qquad (8.8)$$

so

$$\mathbf{w}(n + 1) = (\mathbf{DF} - \mathbf{PK}) \cdot \mathbf{w}(n), \qquad (8.9)$$

and we have a conventional linear stability problem in which we want to choose the vector **K** so that all eigenvalues of the matrix $\mathbf{DF} - \mathbf{PK}$ are within the unit circle. This gives the eigenvalue problem

$$\det(\lambda \mathcal{I} - \mathbf{DF} + \mathbf{PK}) = 0, \qquad (8.10)$$

and in the eigenbasis of the matrix $\mathbf{DF}$, the matrix is

$$\left(\begin{array}{cc} \lambda - \lambda_u + P_u K_u & P_u K_s \\ P_s K_u & \lambda - \lambda_s + P_s K_s \end{array} \right). \tag{8.11}$$

There are clearly many choices of the components (K_s, K_u) of the vector $\mathbf{K}$ which guarantee that the eigenvalues λ lie within the unit circle. Choosing $K_u = \lambda_u / P_u$, the eigenvalues λ are λ_s and zero. This means that choosing the value of $\delta p(n)$ so the projection of $\mathbf{w}(n+1)$ along the unstable manifold is zero, while one among many solutions to this linearized stability problem, is certainly a choice which is likely to be quite robust as one of the eigenvalues, the unstable one, is now set as small as possible.

8.1.1   Targeting

In working with the parameter alterations required to move the orbit onto the stable manifold of a designated unstable periodic orbit we are able to implement the control with great success only after the system moves into the region of the phase space that we wish to control. One has to wait until the dynamics approaches the desired region before control can be initiated. This can often take a long time. It would be very useful to have some technique to persuade the dynamics to quickly approach the control region. In other words we need a net that can be thrown over the phase space in order to capture the trajectory. Once it is in the net we can drag it back to our control cage and lock up its behavior as desired. A method devised for this has been called targeting [SOGY90].

The targeting technique involves both forward and backward iterates of small regions of the phase space. Start as before taking data as given on a surface of section by $\mathbf{v}(n)$, $n = 1, 2, \ldots, N$. Again work in two phase space dimensions, and take the dynamics to be invertible. Suppose the initial point is $\mathbf{v}_i$ and the target point $\mathbf{v}_T$. Without altering the dynamics the time necessary for a typical point on the attractor to enter a ball of size ϵ_T centered on $\mathbf{v}_T$ scales as $\tau \sim 1/\mu(\epsilon_T)$ where μ is the natural measure on the attractor. This measure scales with D_1, the information dimension of the attractor [BP85], so $\tau \sim (1/\epsilon_T)^{D_1}$. Obviously for small ϵ_T the time it takes for a trajectory to wander into the target region can be quite long. This phenomenon is evident in the figure of control of a magnetoelastic ribbon [DRS90] where it took thousands of iterates before the trajectory was near enough to the period one target to implement control. Other examples of even longer duration exist [SOGY90].

Clearly this is not acceptable, and happily it can be improved upon. With $\mathbf{v}(n+1) = \mathbf{F}(\mathbf{v}(n), p)$ the change in one iterate due to a change in the parameter is

$$\delta \mathbf{v}(n+1) = \frac{\partial \mathbf{F}[\mathbf{v}(n), p]}{\partial p} \delta p. \tag{8.12}$$

If we choose δp small, then $\delta\mathbf{v}(n+1)$ sweeps out a small line segment through the point $\mathbf{v}(n+1)$. After m steps governed by the unperturbed map $\mathbf{F}(\mathbf{v},p)$, the line segment will have grown to a length $\delta\mathbf{v}(n+m) \sim \exp[m\lambda_1]\delta\mathbf{v}(n+1)$. Eventually the length will grow to the same order as the size of the attractor, which we call 1. The number of iterates necessary for this is about $-\log[\delta\mathbf{v}(n+1)]/\lambda_1$. Now iterate the target region ϵ_T **backwards** in time. This will transform the target region into an elongated ellipse. The length of the ellipse will grow as $\exp[\lambda_2 m]$. After a total of $-\log[\epsilon_T]/\lambda_2$ iterates the ellipse will have been stretched to the point where its length is the same order as the size of the attractor. λ_2 appears here because we are iterating backwards on a two-dimensional map, and λ_2 is the negative Lyapunov exponent of greatest magnitude.

Generically the line formed by the forward iterates of $\delta\mathbf{v}(n+1)$ will intersect the ellipse formed by the backward iterates of the region ϵ_T. For some value or the parameter in the range $p+\delta p$ the end of the line segment will be *inside* the backward images of ϵ_T. By changing the parameter at time n from the value p to the value p_1 the trajectory will be guaranteed to land inside the target region. The time required for this to happen is

$$t_{\text{Target}} = -\frac{\ln[\delta\mathbf{v}(n+1)]}{\lambda_1} - \frac{\ln[\epsilon_T]}{\lambda_2}, \qquad (8.13)$$

which can be much less than the time given above for the orbit to arrive in the ellipse without additional persuasion.

The targeting strategy above utilizes only one change in the parameter. After the initial perturbation, the parameter value is returned to p. It also assumes that one has complete knowledge of the dynamics. In the presence of noise in or incomplete knowledge of the dynamics the procedure will not work without modifications. The modifications are very small. Instead of only perturbing the dynamics once at time n one should perturb the dynamics at each iterate though the surface. Therefore, the entire forward-backward calculation should be done at $\mathbf{v}(n)$, $\mathbf{v}(n+1)$, $\mathbf{v}(n+2)$, etc. until the target region is reached.

8.2   Other Controls

Instead of using external forces to alter the system parameters one can directly influence the dynamical variables in many cases. In our examples from nonlinear circuits one can feed in voltages or currents into various ports of the circuit and by this direct the dynamics in desired ways [Pyr92, RTA94]

A more or less general description of this involves a driving system for dynamics $\mathbf{d}(t)$ and a response system $\mathbf{x}(t)$, which is the unperturbed dynamics, taken to satisfy

$$\frac{d\mathbf{x}(t)}{dt} = \mathbf{F}(\mathbf{x}(t)). \tag{8.14}$$

There is no bound to the number of ways in which one may couple the drive $\mathbf{d}(t)$ into the response, but the one which has been systematically explored is

$$\frac{d\mathbf{x}(t)}{dt} = \mathbf{F}(\mathbf{x}(t)) - \mathbf{G} \cdot [\mathbf{x}(t) - \mathbf{d}(t)], \tag{8.15}$$

with $\mathbf{G}$ a coupling matrix to be specified. Without detailed analysis we can see that if the eigenvalues of $\mathbf{G}$ are positive and of large enough magnitude, then when $\mathbf{d}(t)$ is any solution of the original dynamics

$$\frac{d\mathbf{d}(t)}{dt} = \mathbf{F}(\mathbf{d}(t)), \tag{8.16}$$

stable or unstable, chaotic or regular, any $\mathbf{x}(t)$ will be driven to it. This happens because the coupling term is vanishing when $\mathbf{d}(t) \approx \mathbf{x}(t)$ leaving $\mathbf{d}(t)$ to satisfy the undriven dynamics. If we know an unstable periodic orbit $\mathbf{x}_P(t)$ of the undriven dynamics and set $\mathbf{d}(t) = \mathbf{x}_P(t)$, then with the eigenvalues of $\mathbf{G}$ large and positive, any solution $\mathbf{x}(t)$, chaotic or not, will be driven to $\mathbf{x}_P(t)$ which will now be stable. This stability is determined by the effective Jacobian matrix $\mathbf{DF} - \mathbf{G}$ whose eigenvalues can all be negative now. For intermediate values of the eigenvalues of $\mathbf{G}$, there may be new stable solutions to the driven dynamics which depend functionally on the drive $\mathbf{d}(t)$.

This approach to control is both quite effective and quite well documented in the standard control literature. It does not rely on any details of the nonlinear features of the $\mathbf{x}(t)$ dynamics, except perhaps to know that unstable periodic orbits abound and can be stabilized this way. it does not require detailed information about the stable and unstable manifolds of the vector field. In that respect it is likely to be quite robust and widely applicable to a variety of dynamical systems, but it also falls out of the main subject of this book.

8.3 Examples of Control

8.3.1 Magnetoelastic Ribbon

The earliest demonstration of the parametric control of chaotic motion was by a group at the Naval Surface Warfare Center [DRS90]. The experiment consisted of a vertically oriented magnetoelastic ribbon placed within a vertical magnetic field $H(t) = H_0 + H_1 \cos(2\pi f t)$. Without the field the ribbon simply buckles under gravity, but since it has a magnetic field dependent Young's modulus, it stiffens with the application of the magnetic

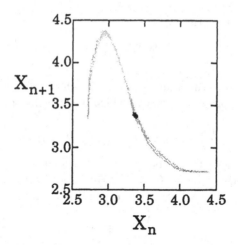

FIGURE 8.2.   Phase space portrait of data from the vibrating magnetoelastic ribbon [DRS90]. The displacement of the ribbon at time $n + 1$ is plotted against the displacement at time n. The dynamics is nearly one dimensional in these data.

field, and the competition between the two forces causes the ribbon to fluctuate. With parameter settings $H_0 = 2.05$ Oe and $H_1 = 0.112$ Oe, chaotic motion is observed in this system with $f = 0.85$ Hz. The data consist of the output voltage of an optical sensor which measured the curvature of the ribbon at a location near its base. This voltage was sampled every period of the magnetic field drive, so $\tau_s = 1.18$ s. The phase space portrait of the data $X(n) = X_n = \text{Voltage}(t_0 + n\tau_s)$ is seen in Figure 8.2 where we plot $X(n) = X_n$ versus $X(n + 1) = X_{n+1}$ to be nearly a one-dimensional map. A fixed point of this map was located at $X_F = 3.398 \pm 0.002$, and a period two orbit at $X_+ = 3.926 \pm 0.004$ and $X_- = 2.870 \pm 0.004$ were located by analysis of the data.

In Figure 8.3 we have the experimental results for a parametric control involving small changes in H_0 of order ± 0.01 Oe for the control of the natural chaos to the fixed point. The experiment was run for as long as 200,000 τ_s, over 65 hours, with results as shown. The control was started at $t = t_0 + 2350\tau_s$ where it is clear that the natural chaotic motion was quickly driven to the desired (previously) unstable fixed point. In Figure 8.4 is a demonstration of the ability to switch from the fixed point to the period two orbit and back. Experiments have also demonstrated the ability to stabilize higher order periodic orbits in this system.

8.3.2   Nonlinear Electric Circuits; Chaotic Lasers

The control of unstable periodic orbits in nonlinear electronic circuits is an interesting application of the parametric control method. A striking example because of its demonstration of versatility in stabilizing high order

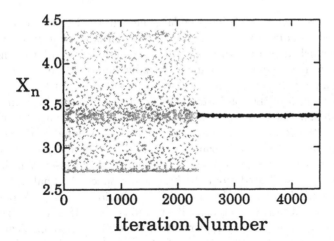

FIGURE 8.3. A time series for the vibration of the magnetoelastic ribbon in the experiment of Ditto, Rauseo, and Spano [DRS90] when the system is driven to the previously unstable fixed point.

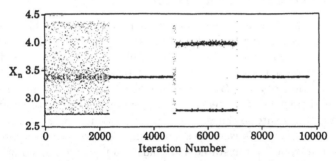

FIGURE 8.4. A time series for the vibration of the magnetoelastic ribbon in the experiment of Ditto, Rauseo, and Spano [DRS90] when the system is driven over the course of the experiment to the previously unstable fixed point to the previously unstable period two orbit—all under the control of the experimenters.

periodic orbits is that of Hunt [Hun91] to a circuit with a solid state diode resonator. The diode is driven at 53 kHz and acts as the control with small variations in this voltage made according the rule: when the peak current is within a certain window around a desired value, the drive voltage is modulated proportional to the difference between the observed peak current and the desired peak current. The changes in drive voltage are as small as 0.5% for stabilizing a fixed point while it may rise to as high as 9% for control of a period 21 orbit. These are relatively small parameter variations.

This method, called occasional proportional feedback, seems to be essentially that described above for the control of low order orbits as the variation in the parameter is so small, but when the stabilization of high order periodic orbits is achieved, the magnitude of the parameter variation, while still relatively small, may be large enough to alter the attractor and introduce periodic orbits where no unstable periodic orbit was present for the unperturbed system. In this regard the method is similar to the direct manipulation of a system dynamical variable described before.

This same direct control method was utilized to stabilize infrared intensity fluctuations of the solid state laser described in some detail in Chapter 11 [RJM*92]. Using the power into the laser diode driving the Nd:YAG system as the control parameter, occasional proportional feedback as just described allowed the stabilization of period one, period four, and period nine orbits. An important feature of the control of this laser is in the power which was redistributed from the broadband spectral signal of the chaos into the period one, that is sinusoidal, oscillations. The peak power under the peak of infrared intensity fluctuations near 118 kHz was increased by a factor of nearly 15 by implementing the control scheme. Clearly when spectrally sharply defined variation of a light signal is desirable, the method promises clear technological possibilities.

8.3.3 Control in Cardiac Tissue

The parametric control scheme has been implemented in biological systems as well. A striking example of this is its application to the control of chaotic motions in a piece of rabbit heart placed in an oxygenated Kreb's solution and pulsed with a square wave with amplitude of 10 to 30 volts and plateau time of 3 ms. By adding a ouabain-epinephrine mixture to the rabbit heart preparation, irregular response was induced with an apparent period doubling route to chaos. The measured quantity was the interval between beats, and in this variable a time delay phase space was constructed in which the stable and unstable manifolds of a flip saddle fixed point were identified. This fixed point appears as a short interval between beats followed by a long interval. The control consisted of shortening the interval between beats by pulsing the cardiac preparation electrically to shorten a selected interval between beats. It was not possible to lengthen the interval with this intervention. Further, since this was a laboratory experiment on real material, it was never possible to precisely implement the driving of the system to the stable manifold as dictated by the pure parametric control rule. The system was driven quite close to the fixed point, and that proved quit enough to move the dynamics from chaotic to evolution along a period 3 orbit in 8 out of the 11 samples tried. This near approach to the desired fixed point is also inherent in the experiments on an NMR laser [RFBB93] which was successful in establishing control.

8.3.4 Experimental Targeting

Using the same magnetoelastic ribbon experiment a demonstration of targeting was performed. Only the forward portion of the targeting method was used. This is equivalent to setting $m_2 = 0$ and waiting for the stretched line segment to contact ϵ_T. The results indicate that for the magnetoelastic ribbon experiment without targeting it required about 500 iterates to reach a typically chosen ϵ_T. With targeting the number of iterates typically was reduced to 20 [SDG*92]. The power of using the nonlinear dynamics of the process is quite apparent.

8.4 A Few (Irreverent) Remarks About Chaos and Control

The remarkable success achieved in being able to reduce chaotic motion to periodic motion by changing the dynamical system albeit slightly with the addition of new degrees of freedom (time dependent parameters or external forces) is quite impressive. There is no doubt that reducing the problem to a previously solved problem is sure to have numerous technological applications and will become an important tool for engineering design when chaotic motions are unwanted. One can be sure that many new methods built on the basic ideas of parametric control or carefully selected external forcing will be developed.

It is likely to be a minority opinion that this thrust misses the new capabilities available from the discovery of chaos. Since chaos is a signal intermediate between regular motions and high dimensional noise, perhaps one should be looking to seeking applications of this new class of signals which explore wide regions of phase space in a structured fashion. If I had an excellent example, this point of view might be more persuasive, but at the moment it is just a point of view. It is natural when faced with a new phenomenon to try to reduce it to previously understood matters, and it is often very successful both scientifically and technically. We should hesitate to pass by the opportunities that new degrees of freedom must certainly present us and fail to stray from periodic motions as the goal of working systems. As suggested in the introduction, if biological systems had adopted periodic or quasi-periodic motions, they might have failed to evolve.

9

Synchronization
of Chaotic Systems

Synchronization among dynamical variables in coupled chaotic systems would appear to be almost an oxymoron as the definition of chaos includes the rapid decorrelation of nearby orbits due to the instabilities throughout phase space. Nonetheless [FY83, AVR86, PC90], it is quite straightforward to establish situations where synchronization occurs between signals, each of which is chaotic. This synchronization has clear applications to communications [VR93] and control [Pyr92, RTA94]. Further, it may be responsible for the saturation of the invariant characteristics of chaos in chains of coupled nonlinear oscillators and in more complicated systems [GRS84, WR90].

The apparent puzzle associated with synchronizing coupled nonlinear systems each of which may operate in chaotic motion may be resolved in a qualitative sense if one thinks about it as arising within a very special range of parameter settings where the joint phase space of the two systems is stable to motion on a subspace where the motion may be regular or another variety of chaotic evolution.

Just as with our discussion of control in Chapter 8 we stray from the main theme of this monograph in taking up the topic of synchronization. In the last section of this chapter, however, we shall make contact with the mainstream of this book by addressing matters related to the determination of the synchronization of two signals from dissimilar systems where we have incomplete knowledge of the systems and have only observed signals from each. In that situation we will require our notions of reconstruction of state space, false nearest neighbors, and the like. In dealing with that issue we shall explicitly recognize that synchronization, in any practical sense, requires examination of the predictability of one signal from another, and for that we have well developed tools as the reader has seen.

9.1   Identical Systems

There is a sizable and growing literature on the synchronization of nonlinear systems. This is partly due to their intrinsic interest, and it is partly for the clear applicability of the concept. The class of synchronized chaotic oscillations which has been most analyzed is that where two identical systems are coupled and then evolve with corresponding dynamical variables having exactly the same behavior in time. That is, when the corresponding dynamical variables of the two systems are proportional to each other. There are two categories of systems where this sort of behavior is observed.

  i The first category was introduced by Pecora and Carroll [PC90, CP91] and consists of a **driving** system and of an identical **response** system. The response is forced by the driving system with one of its outputs. Each subsystem—driving and response—can exhibit chaotic behavior.

 ii The second category includes coupled systems which at zero coupling are identical to each other, and which each display chaotic behavior [FY83, AVR86, RVR*92, AVPS91]. When the systems are coupled through a combination of differences of their dynamical variables in a manner which introduces additional dissipation into the overall dynamics, the systems may demonstrate identical oscillations associated with the onset of synchronization. This means the corresponding dynamical variables in each subsystem are tracing out precisely the same points in state space after a possible transient.

The region in parameter space where this identity between driving and response variables is observed need not be large.

The first type of synchronization identifies two identical dynamical systems $\mathbf{x}_1 \rightarrow \mathbf{F}(\mathbf{x}_1)$ and $\mathbf{x}_2 \rightarrow \mathbf{F}(\mathbf{x}_2)$. One (or more) of the dynamical variables from system one is then used as a replacement for the corresponding dynamical variables in system two. To achieve this we divide each system into two parts in which one subpart will be taken by an appropriate communications link to the second system. There appear to be no a priori rules for making this division, and the criterion given below for synchronization of the two systems using a given division is clearly a posteriori and empirical.

To achieve the desired division we write for system one $\mathbf{x}_1(t) = [\mathbf{y}_1(t), z_1(t)]$ where $\mathbf{y}_1(t)$ is a dynamical variable with dimension less than that of $\mathbf{x}_1(t)$, and $\mathbf{z}(t)$ is m dimensional. For simplicity we will always take $m = 1$. $z(t)$ is the part of the overall system $\mathbf{x}_1$ which will be carried over to system two. The equations of motion of system one are now

$$\frac{d\mathbf{y}_1(t)}{dt} = \mathbf{F}_y(\mathbf{y}_1(t), z(t)),$$

$$\frac{dz(t)}{dt} = F_z(\mathbf{y}_1(t), z(t)). \qquad (9.1)$$

Now the signal $z(t)$ is used to drive system two which we divide as $\mathbf{x}_2(t) = [\mathbf{y}_2(t), z(t)]$ so the last component of the $\mathbf{x}_2$ system is now identical to the last component of the $\mathbf{x}_1$ system. The equations of motion for the $\mathbf{y}_2(t)$ become

$$\frac{d\mathbf{y}_2(t)}{dt} = \mathbf{F}_y(\mathbf{y}_2(t), z(t)), \qquad (9.2)$$

and there is no equation required for $z(t)$ since it is prescribed by the evolution of the drive system. The $\mathbf{y}_2$ dynamics is called the response.

Synchronization is said to occur when $\mathbf{y}_1(t) = \mathbf{y}_2(t)$ for a wide range of initial conditions, namely, the basin of attraction for synchronization. This condition is stable when small deviations $\Delta\mathbf{y}(t) \equiv \mathbf{y}_1(t) - \mathbf{y}_2(t)$ go to zero in time. If $\Delta\mathbf{y}(t)$ is small and remains small, its dynamics are approximately

$$\frac{d\Delta\mathbf{y}(t)}{dt} = \mathbf{DF}_y(\mathbf{y}(t), z(t)) \cdot \Delta\mathbf{y}(t), \qquad (9.3)$$

where $\mathbf{DF}_y(\bullet, z(t))$ is the $d-1$-dimensional **conditional** Jacobian around either $\mathbf{y}_1$ or $\mathbf{y}_2$, and $\mathbf{y}$ can be either of these. Conditional means conditioned on the given value of $z(t)$. If the conditional Lyapunov exponents of the conditional Oseledec matrix made out of $\mathbf{DF}_y(\bullet, z(t))$ are all negative, then synchronization occurs. This means that for any $\Delta\mathbf{y}(0), \Delta\mathbf{y}(t) \to 0$ for t large.

Using the familiar Lorenz model as an example, we list in Table 9.1 the values of the conditional Lyapunov exponents when one selects as the drive first $x(t)$ then $y(t)$ and finally $z(t)$ from one Lorenz system coupled to another with identical parameters σ, b, and r.

From this table we can see that using either the x or y dynamical variables as the drive, the (y, z) or the (x, z) subsystems, respectively, of two Lorenz systems will synchronize. If we were to choose z as the drive, the (x, y) subsystems would not synchronize. In Figure 9.1 we plot $y_{one}(t) - y_{two}(t)$ and $z_{one}(t) - z_{two}(t)$ when $x(t)$ is used as the synchronizing drive. In Figure 9.2 we plot $x_{one}(t) - x_{two}(t)$ and $z_{one}(t) - z_{two}(t)$ when y is used as

Lorenz System	Drive	Response	Conditional Exponents
$\sigma = 16.0$	x	(y, z)	$(-2.50, -2.50)$
$b = 4.0$	y	(x, z)	$(-3.95, -16.0)$
$r = 45.92$	z	(x, y)	$(7.89 \times 10^{-3}, -17.0)$

TABLE 9.1.

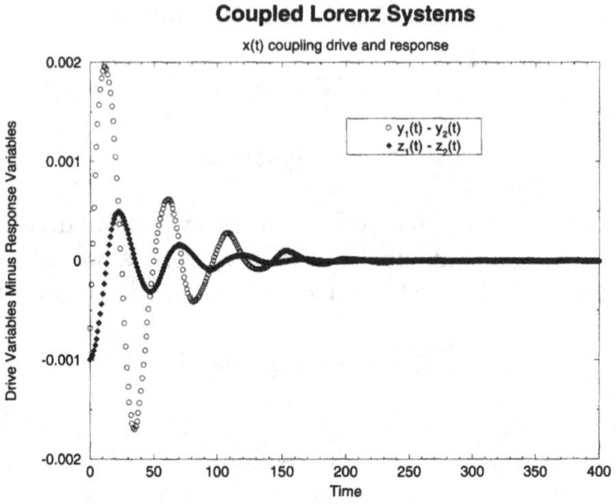

Coupled Lorenz Systems

x(t) coupling drive and response

FIGURE 9.1.   Two identical Lorenz systems Equation (3.5) are coupled through their $x(t)$ dynamical variable using the method of Pecora and Carroll [PC90]. We show the difference between the $y(t)$ variables and the $z(t)$ variables for the two systems coupled this way. The conditional Lyapunov exponents as shown in Table 9.1 for this coupling are all negative, and the coupled system synchronizes by having the $y(t)$ dynamical variables and the $z(t)$ dynamical variables become equal after some transients related to initial conditions.

the synchronizing drive. As the conditional Lyapunov exponents are negative for these cases, we anticipate and see synchronization of the subsystems. Finally in Figure 9.3 we plot $x_{\mathrm{one}}(t) - x_{\mathrm{two}}(t)$ and $y_{\mathrm{one}}(t) - y_{\mathrm{two}}(t)$ when z is used as the synchronizing drive. This last case has a positive conditional Lyapunov exponent, and the lack of synchronization shown in Figure 9.3 is indicative of that. These results demonstrate clearly what synchronization by this coupling of dynamical variables between identical systems means. Pecora and Carroll [PC90] also explore the sensitivity of this synchronization on variations in the parameters in the vector field $\mathbf{F}(\bullet)$. For the Lorenz dynamics, the synchronization is rather robust against this parametric variation. Basically this means there is a range of parameters where the conditional Lyapunov exponents are all negative.

The second class of coupling connecting similar chaotic systems comes in an example from Afraimovich, Verichev, and Rabinovich [AVR86]. The example considered is the following

$$\frac{dx_1(t)}{dt} = y_1(t),$$

$$\frac{dy_1(t)}{dt} = -ky_1(t) - x_1(t)(1 + q\cos[\theta(t)]) + x_1(t)^2) - c(y_1(t) - y_2(t)),$$

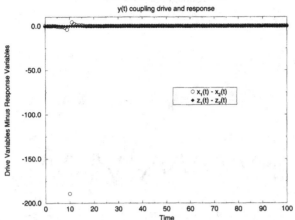

FIGURE 9.2.   Two identical Lorenz systems Equation (3.5) are coupled through their $y(t)$ dynamical variable using the method of Pecora and Carroll [PC90]. We show the difference between the $x(t)$ variables and the $z(t)$ variables for the two systems coupled this way. The conditional Lyapunov exponents as shown in Table 10.1 for this coupling are all negative, and the coupled system synchronizes by having the $x(t)$ dynamical variables and the $z(t)$ dynamical variables become equal after some transients related to initial conditions.

$$\frac{dx_2(t)}{dt} = y_2(t),$$

$$\frac{dy_2(t)}{dt} = -ky_2(t) - x_2(t)(1 + q\cos[\theta(t)]) + x_2(t)^2) - c(y_2(t) - y_1(t)),$$

$$\frac{d\theta(t)}{dt} = \Omega. \tag{9.4}$$

This coupling for $c > 0$ is a dissipative kind of interaction for these systems. Increasing c from zero effectively increases the damping k in each system and drive the systems to $y_1(t) = y_2(t)$ even if each system is undergoing chaotic motions. Deviations from synchronization are quickly damped and the coupling dynamics settles into identical motion in each subsystem.

When the coupling $c = 0$ each of the oscillators shows chaotic evolution over a wide range of the parameters k, q, and Ω. In Figure 9.4 we show the regular oscillations of this system for $c = 0$ when $q = 50.0, \Omega = 2.0$, and $k = 0.50$. We display the motion in the (x, y) plane. In Figure 9.5, this is shown for $k = 0.45$. When we couple the oscillators the two subsystems come into synchrony as c is increased. In Figure 9.6 we plot the RMS value of the difference $x_1(t) - x_2(t)$ as a function of c for $q = 50.0, \Omega = 2.0$, and $k = 0.45$. For reassurance that we are synchronizing chaotic motions, we

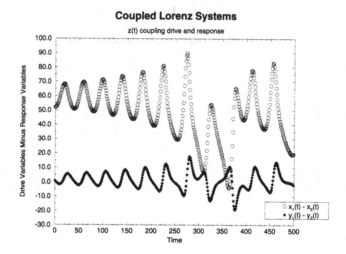

FIGURE 9.3.   Two identical Lorenz systems Equation (3.5) are coupled through their $z(t)$ dynamical variable using the method of Pecora and Carroll [PC90]. We show the difference between the $x(t)$ variables and the $y(t)$ variables for the two systems coupled this way. One of the conditional Lyapunov exponents for this coupling is positive as shown in Table 10.1, and the coupled system fails to synchronize, as we can see.

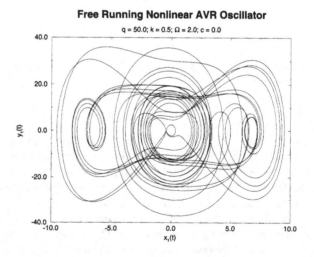

FIGURE 9.4.   The regular attractor associated with the equations of AVR (Afraimovich, Verichev, and Rabinovich) Equation (9.4) for one of the oscillators when the coupling is off: $c = 0.0$, while $q = 50.0$, $k = 0.5$, and $\Omega = 2.0$. This is projected onto the $(x_1(t), y_1(t))$ plane.

Free Running Nonlinear AVR Oscillator

q = 50.0; k = 0.45; Ω = 2.0; c = 0.0

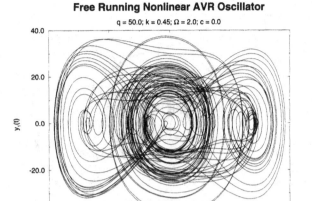

FIGURE 9.5. The irregular attractor associated with the equations of AVR (Afraimovich, Verichev, and Rabinovich) Equation (9.4) for one of the oscillators when the coupling is off: $c = 0.0$, while $q = 50.0, k = 0.45$, and $\Omega = 2.0$. This is projected onto the $(x_1(t), y_1(t))$ plane.

Synchronization of Nonlinear Oscillators

AVR Oscillators; k = 0.45; q = 50.0; Ω = 2.0

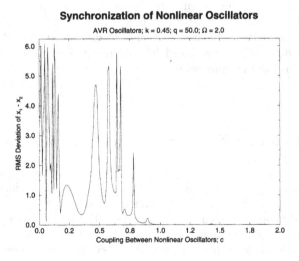

FIGURE 9.6. The synchronization of the AVR system (Afraimovich, Verichev, and Rabinovich) Equation (9.4) as a function of the coupling parameter c in the range $0.0 \leq c \leq 2.0$. The system parameters are $k = 0.45, q = 50.0$, and $\Omega = 2.0$ for each oscillator. Synchronization also occurs when the oscillators are not identical.

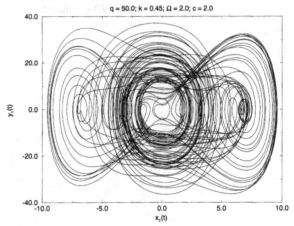

FIGURE 9.7.   The projection onto the $(x_1(t), y_1(t))$ plane of the motion of oscillator one in the AVR system (Afraimovich, Verichev, and Rabinovich) Equation (9.4) when the coupling between the systems is nonzero: $c = 2.0$. This is to assure us that the motion is chaotic when $c = 2.0$, and the systems are coupled.

show in Figure 9.7 the projection of the orbit onto the (x_1, y_1) plane at the value $c = 2.0$ where the systems are clearly synchronized.

It seems clear that this kind of coupling between identical chaotic systems can synchronize their dynamical variables and the source of this is the additional dissipation introduced into each equation when the variables are not following the same orbits.

9.2   Dissimilar Systems

In the coupling of dissimilar systems one has interesting possibilities of synchronization among the variables of one subsystem with the other which do not appear natural when the systems are identical. Suppose we have two dissimilar dynamical systems $\mathbf{x}_1(t) \to \mathbf{F}_1(\mathbf{x}_1(t))$ and $\mathbf{x}_2(t) \to \mathbf{F}_2(\mathbf{x}_2(t))$ and wish to know if some subset of variables $\mathbf{y}_1(t)$ of system one and another subset of variables $\mathbf{y}_2(t)$ from system two are synchronized. This means that within the full dynamical space of variables $(\mathbf{y}_1, \mathbf{y}_2)$ there is a relationship $\mathbf{y}_1(t) = \boldsymbol{\phi}(\mathbf{y}_2(t))$ implying that

    i the motion is restricted to a lower dimensional manifold than the full space, and even more important, ii the evolution of the $\mathbf{y}_1(t)$ is *predictable* in terms of the $\mathbf{y}_2(t)$ and vice versa if the function $\phi(\bullet)$ is invertible. The usual sense of synchronization is confined to choosing $\phi(\bullet)$ to be the identity.

Suppose all we know about two coupled systems is a time series of measurements of one the variables from each, say $u_1(t)$ from the $\mathbf{x}_1$ and $u_2(t)$ from the $\mathbf{x}_2$. How will we know if these are synchronized if it happens that $u_1(t) \neq u_2(t)$, which should be the most common occurrence? What we need to do is develop some tests, useful in reconstructed phase space, for each of the dynamical systems which allows us to determine if one is predictable in terms of the other. This we must be able to do without knowledge of the equations of motion for either system. We now give two different ways to examine this question of predictability and thus synchronization in this general sort of situation.

First let us illustrate this with an artificial example. We consider two coupled Rössler [Roe76] oscillators which are coupled in a unidirectional fashion as indicated by

$$\text{driving system} \quad : \quad \begin{cases} \frac{dx_1(t)}{dt} = -(x_2(t) + x_3(t)), \\ \frac{dx_2(t)}{dt} = x_1(t) + 0.2x_2(t), \\ \frac{dx_3(t)}{dt} = 0.2 + x_3(t)(x_1(t) - \mu), \end{cases} \quad (9.5)$$

$$\text{response system} \quad : \quad \begin{cases} \frac{dy_1(t)}{dt} = -(y_2(t) + y_3(t)) \\ \qquad\quad - g(y_1(t) - x_1(t)), \\ \frac{dy_2(t)}{dt} = y_1(t) + 0.2y_2(t), \\ \frac{dy_3(t)}{dt} = 0.2 + y_3(t)(y_1(t) - \mu) \end{cases} \quad (9.6)$$

with $\mu = 5.7$.

The subspace

$$x_1(t) = y_1(t), \qquad x_2(t) = y_2(t), \qquad x_3(t) = y_3(t) \qquad (9.7)$$

of the full $(x_1, x_2, x_3, y_1, y_2, y_3)$ space defines synchronized motions. This is stable at $g = 0.20$ and unstable at $g = 0.15$. To confirm this we computed the conditional Lyapunov exponents [PC90, RVR*92] associated with the behavior of the response system for these two values of g, conditioned on the value of $x_1(t)$. The largest conditional Lyapunov exponent is -0.021 at $g = 0.20$ but becomes $+0.024$ at $g = 0.15$.

Suppose we know only the variables x_2 and y_2 of these systems Equation (9.5) and Equation (9.6). Then a plot of $y_2(t)$ versus $x_2(t)$ will indicate the onset of the standard kind of synchronization, namely, $y_2(t) = x_2(t)$. For $g = 0.20$, shown in Figure 9.8, we see a sharp straight line, as it should be for these synchronized oscillations as in Equation (9.7). Figure 9.9 shows the same plot for $g = 0.15$. This corresponds to unsynchronized motions in the coupled systems.

Now we construct a response system which exhibits a *generalized* kind of synchronization [RSTA95]. We make the simple nonlinear transformation

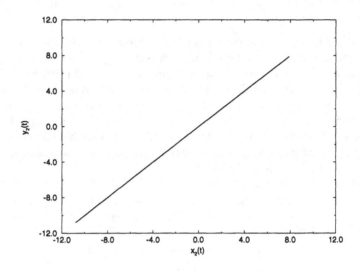

FIGURE 9.8.   The projection of the chaotic attractor generated by the system
Equation (9.5) and Equation (9.6) onto the plane (x_2, y_2) in synchronized behavior.
Here $g = 0.2$.

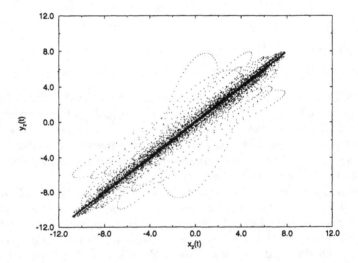

FIGURE 9.9.   The projection of the chaotic attractor generated by the system Equation (9.5) and Equation (9.6) onto the plane (x_2, y_2) in unsynchronized
behavior. Here $g = 0.15$.

among the response variables

$$z_1(t) = y_1(t), \qquad z_2(t) = y_2(t) + ay_3(t) + by_3(t)^2, \qquad z_3(t) = y_3(t). \quad (9.8)$$

We take $a = 0.4$ and $b = -0.008$. This system should have the same properties as the original. In the $z_a(t)$ variables the equations read

$$\frac{dz_1(t)}{dt} = -(z_2(t) + (1-a)z_3(t) - bz_3(t)^2) - g(z_1(t) - x_1(t)),$$

$$\frac{dz_2(t)}{dt} = z_1(t) + 0.2(z_2(t) - az_3(t) - bz_3(t)^2),$$

$$+(a + 2bz_3(t))(0.2 + z_3(t)(z_1(t) - \mu)),$$

$$\frac{dz_3(t)}{dt} = 0.2 + z_3(t)(z_1(t) - \mu). \quad (9.9)$$

When we plot $z_2(t)$ versus $x_2(t)$ for the synchronized state at $g = 0.20$, we no longer see a straight line but the more complex object in Figure 9.10. The plot of $z_2(t)$ vs $x_2(t)$ for the synchronized state looks "cloudy" or "fuzzy." Nonetheless, since all we did was perform a smooth change of coordinates, we know that synchronization, as a basic property of these coupled oscillators, cannot have been lost. A similar plot for $g = 0.15$ which corresponds to unsynchronized motions is shown in Figure 9.11. We now seek ways in which to distinguish these two states of the system, synchronized and unsynchronized, using observations of the variables $x_2(t)$ and $z_2(t)$ alone. This corresponds to the type of situation we might well encounter in determining whether two observed dynamical oscillators are synchronized or not.

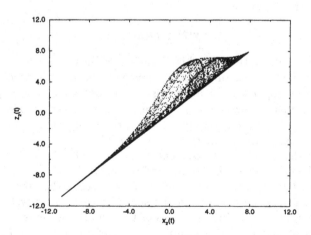

FIGURE 9.10. The projection of the chaotic attractor generated by the system Equation (9.5) and Equation (9.9) on the plane (x_2, z_2) in synchronized behavior, $g = 0.2$.

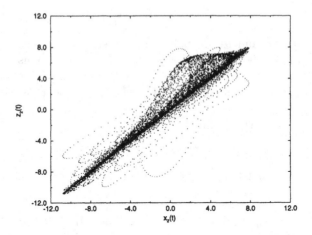

FIGURE 9.11.   The projection of the chaotic attractor generated by the system Equation (9.5)and Equation (9.9) on the plane (x_2, z_2) in unsynchronized behavior, $g = 0.15$.

9.3   Mutual False Nearest Neighbors

By making changes of coordinates we demonstrated that simply looking for synchronization as the identity matching of drive and response variables $\mathbf{y}(t) = \mathbf{x}(t)$ will not uncover synchronization in general settings. For this we require an algorithm which rests on the existence of the generalized synchronization condition

$$\mathbf{y}(t) = \boldsymbol{\phi}(\mathbf{x}(t)). \qquad (9.10)$$

We have emphasized that it is this condition among drive variables $\mathbf{x}(t)$ and response variables $\mathbf{y}(t)$ which is the essence of synchronization of a driving system and a response system. In this section we explore one requirement imposed by this connection, and in the next section we examine another. In our concluding section we will mention others we have not yet explored.

The keystone of the characteristic of coupled systems which we are considering in this section is the concept of local neighborliness. When trajectories in the phase spaces of driving and response systems are connected by $\mathbf{y}(t) = \boldsymbol{\phi}(\mathbf{x}(t))$, two *close* states in the phase space of the response system correspond to two *close* states in the space of the driving system. Let us consider a set of points in the spaces of the coupled systems coming from finite segments of trajectories sampled at moments of time $t_n = t_{\text{sampling}} \cdot n$ where n is integer. Pick an arbitrary point $\mathbf{x}_n = \mathbf{x}(t_n)$ in the phase space of the driving system. Suppose the nearest phase space neighbor of this point has time index n_{NND}. Then as long as the trajectories are connected by the relation (9.10), the point $\mathbf{y}_n$ in the space of the response system will have point $\mathbf{y}_{n_{NND}}$, as a close neighbor.

In a few words what we are looking for is a geometric connection between the driving and response systems which preserves identity of neighborhoods in state space. This is a kind of correlation between observed dynamical variables one from the driving system and one from the response system. It is not a linear correlation, of course, as the underlying dynamics is nonlinear. We could cast the statistical tests we are about to consider in terms of correlations between response and driving variables. If we were to do so, we would probably prefer to seek a nonlinear correlation among these variables using a statistic such as mutual information. This is quite computationally intensive, and for our purposes we have found the statistics below to be adequate, if not fully nonlinear. Indeed they emphasize the geometry more than the nonlinearity of the dynamics.

This property can be characterized by a numerical parameter. To form this parameter we notice that when Equation (9.10) holds and the distances between two nearest neighbors in the phase spaces of the driving and response systems are small, we can write

$$
\begin{aligned}
\mathbf{y}_n - \mathbf{y}_{n_{NND}} &= \boldsymbol{\phi}(\mathbf{x}_n) - \boldsymbol{\phi}(\mathbf{x}_{n_{NND}}), \\
&\approx \mathbf{D}\boldsymbol{\phi}(\mathbf{x}_n) \cdot (\mathbf{x}_n - \mathbf{x}_{n_{NND}}),
\end{aligned}
\tag{9.11}
$$

where $\mathbf{D}\boldsymbol{\phi}(\mathbf{x}_n)$ is the Jacobian matrix of the transformation $\boldsymbol{\phi}$ evaluated at location $\mathbf{x}_n$. Similarly we go to time index "n" and observe the response vector $\mathbf{y}_n$ and locate its nearest neighbor $\mathbf{y}_{n_{NNR}}$ which comes at time index n_{NNR}. Again, when $\mathbf{y}(t) = \boldsymbol{\phi}(\mathbf{x}(t))$ we can write

$$
\begin{aligned}
\mathbf{y}_n - \mathbf{y}_{n_{NNR}} &= \boldsymbol{\phi}(\mathbf{x}_n) - \boldsymbol{\phi}(\mathbf{x}_{n_{NNR}}) \\
&\approx \mathbf{D}\boldsymbol{\phi}(\mathbf{x}_n) \cdot (\mathbf{x}_n - \mathbf{x}_{n_{NNR}}).
\end{aligned}
\tag{9.12}
$$

This suggests that the ratio

$$
\frac{|\mathbf{y}_n - \mathbf{y}_{n_{NND}}|}{|\mathbf{x}_n - \mathbf{x}_{n_{NND}}|} \frac{|\mathbf{x}_n - \mathbf{x}_{n_{NNR}}|}{|\mathbf{y}_n - \mathbf{y}_{n_{NNR}}|},
\tag{9.13}
$$

which we call the *mutual false nearest neighbors* (MFNN) parameter should be of order of unity when the systems are synchronized in the sense that $\mathbf{y}(t) = \boldsymbol{\phi}(\mathbf{x}(t))$. Generally, if the synchronization relation $\mathbf{y}(t) = \boldsymbol{\phi}(\mathbf{x}(t))$ does not hold, then this parameter should on average be of order (size of attractor squared)/(distance between nearest neighbors squared).

In experiments, we usually do not have the luxury of working with the actual vectors of phase space variables. Normally only the time series of a single variable is available to characterize the behavior of each system. Therefore, to be able to analyze experimental as well as numerical data we will rely upon the phase space reconstruction method.

Suppose we have observed a scalar time series $r(n)$ of some variable in the response system. We form the response vector

$$\mathbf{r}(n) = [r(n), r(n - T_r), r(n - 2T_r), \ldots, r(n - (d_r - 1)T_r)], \qquad (9.14)$$

where T_r is an integer multiple of the sampling time for the observations of the response system. T_r is selected by average mutual information. We call the d_r-dimensional space associated with these vectors $\mathbf{R}_E$. We also observe a scalar time series $d(t)$ of a variable from the driving system and form the vector

$$\mathbf{d}(m) = [d(m), d(m - T_d), d(m - 2T_d), \ldots, d(m - (d_d - 1)T_d)], \qquad (9.15)$$

where T_d is an integer multiple of the sampling time for the observations of the driving system. We call the space of these vectors $\mathbf{D}_E$. d_r and d_d are each larger than the respective global embedding dimensions required to unfold the response and the driving attractors, respectively.

By Takens' theorem, the attractor in embedding space $\mathbf{D}_E$ inherits all properties of the attractor of the driving system. Similarly, when the coupling is not zero, the attractor in the embedding space $\mathbf{R}_E$ is equivalent to the attractor of whole system: *driving+response*. When there is a transformation ϕ that relates the trajectories in the subspaces of the driving and response systems, there must also be a transformation that relates the trajectory in the phase space of the driving system to that in the phase space of the combined system *driving+response*. Therefore the conclusions made above concerning the properties of nearest neighbors for synchronized and unsynchronized behaviors are still true when they are applied to the reconstructed spaces $\mathbf{D}_E$ and $\mathbf{R}_E$. At the same time, using the language of embedding space reconstruction, we can give now a different interpretation to this property. A criterion similar to the one we use here forms the basis of the false neighbors method for determining the minimum dimension that is required to unfold the attractor without self-crossing of the trajectories when one uses embedding methods to reconstruct the attractor from time series. A similar picture is observed when we consider points in the embedding space of the driving signal and points in the embedding space of the response signal. If $\mathbf{r}(t)$ can be obtained by a transformation of $\mathbf{d}(t)$, that is, the response system is synchronized, then the attractors constructed in spaces $\mathbf{R}_E$ and $\mathbf{D}_E$ can be considered as different embeddings of the same attractor, and because of Takens' theorem each provides an equally appropriate set of coordinates. Therefore, if points with time indices n and n_{NND} were nearest neighbors in $\mathbf{D}_E$ space they will be close neighbors in $\mathbf{R}_E$ space too.

Thus we see that the points in the time delay embedding spaces possess the same neighborliness properties as points in the original spaces of physical variables, and we can conclude that the MFNN parameter can be computed according to the ratio (9.13) even when it is computed using the trajectories in embedding spaces of driving and response instead of trajectories in original spaces of $\mathbf{x}$ and $\mathbf{y}$. When the systems are synchronized and d_r and d_d are large enough, the MFNN parameter should be of order one at all locations. When the systems are not synchronized, this parameter will be large at some locations on the attractor.

As stated so far this test has a few problems which are connected with the nontrivial scaling of the MFNN parameter with the embedding dimensions of driving and of response. There are two causes for these problems both of which can make the practical value of this test questionable. First of all, when we use a time series of fixed length the average distance between the nearest neighbors becomes larger and is comparable with the size of attractor in high embedding dimensions. As a result the denominator of the ratio (9.13) grows with increasing embedding dimension. The numerator for *unsynchronized* trajectories is always of the order of the square of the size of the attractor and, therefore, the ratio itself may become small as the embedding dimension increases, even if there is no synchronized motions in the system. On the other hand, when a time series of fixed length is embedded in spaces of higher and higher embedding dimensions, the population density in the spaces decreases. Eventually this leads to loss of any fine structure at high embedding dimensions. The nearest neighbor that is found in high-dimensional space may no longer be a close neighbor. As a result it becomes more and more difficult to uncover a connection between two trajectories when there is such a connection. The MFNN parameter defined by the ratio (9.13) can be large even when the systems are synchronized.

Thus the variation of the parameter with embedding dimensions occurs for two reasons: (1) the search for neighbors is performed in spaces of changing dimensions, and (2) the distances are computed in the spaces of changing dimensions. The former cannot be helped. However, without changing the main idea, we can change the test so that all distances are computed in the same space.

The modified MFNN parameter is constructed as follows. We embed the response time series in the space $\mathbf{R}_E$ of dimension d_r which is then fixed. d_r must not be less than the minimum dimension necessary to unfold the attractor corresponding to the response time series. All distances will be computed in this space. d_d is variable and is larger than the minimum dimension needed to unfold the driver attractor without self-crossing. For each d_d we go to time index "n" and locate the nearest neighbor of point $\mathbf{d}(n)$ which comes at time index n_{NND}. We also find the nearest neighbor $\mathbf{r}(n_{NNR})$ of point $\mathbf{r}(n)$ in the response embedding space. Then form the ratio

$$\frac{|\mathbf{r}(n) - \mathbf{r}(n_{NNR})|^2}{|\mathbf{r}(n) - \mathbf{r}(n_{NND})|^2} \tag{9.16}$$

which is a less symmetric form of the ratio (9.13). We evaluate the squares of Cartesian distances instead of the distances themselves to reduce the computation time. Finally, to compensate for the increase of the MFNN parameter in high dimensions due to the sparseness of the phase space population we divide the ratio (9.16) by the same parameter, computed for the driving time series. This brings us to the final form of the MFFN parameter:

$$P(n, d_r, d_d) = \frac{|\mathbf{d}'(n) - \mathbf{d}'(n_{NND'})|^2}{|\mathbf{d}(n) - \mathbf{d}(n_{NND})|^2} \frac{|\mathbf{r}(n) - \mathbf{r}(n_{NNR})|^2}{|\mathbf{r}(n) - \mathbf{r}(n_{NND})|^2}, \qquad (9.17)$$

where $\mathbf{d}'$ are vectors of driving time series embedded in the space of dimension d_r and $\mathbf{d}'(n_{NDD'})$ is the nearest neighbor of point $\mathbf{d}'(n)$ in this d_r dimensional space. This parameter should on the average be of order of unity for synchronized trajectories. For unsynchronized trajectories we expect $P(n, d_r, d_d) \gg 1$.

It should be understood that $P(n, d_r, d_d)$ is a local characteristic of the attractor in the combined $\mathbf{D}_E \oplus \mathbf{R}_E$ space. To obtain reliable information about the regime in which the system evolves as synchronized or unsynchronized motion, we must determine a statistical ensemble of MFNN values computed at a number of locations on the attractor. Examining the statistical distribution of MFNN is especially important when one studies the system close to the onset of synchronization. In many cases synchronized chaotic motions appear after an intermittent regime in which the system evolves nearly on the manifold of synchronized motions, but leaves it for short periods of time during unsynchronized outbursts. In this case $P(n, d_r, d_d)$ will be of order unity at almost all points but will become large at a small number of points corresponding to these outbursts. Thus it may be instructive to study the histograms of a set of MFNN parameters to obtain the maximum information about the system. However, one can distinguish between synchronized and unsynchronized behaviors by studying the average value $\bar{P}(d_r, d_d)$ of the MFNN parameter alone.

9.3.1   Mutual False Nearest Neighbors; Coupled Roëssler Systems

The mutual false nearest neighbor method is now applied to the directionally coupled Roëssler systems Equation (9.5) and Equation (9.6). In each of the calculations discussed in this section we used a total of 60,000 data points. We employed a fourth-order Runge-Kutta integrator with time step 0.02 to generate the data, and the data was sampled with time step 0.8. The time delays T_r and T_d were each taken to be 5. The embedding dimension $d_r = 4$ was selected, according to the method described in, and only d_d was varied in our calculations. We computed $P(n, d_r, d_d)$ at 10,000 different locations on the attractor and used them to evaluate the average values $\bar{P}(d_r, d_d)$ of the MFNN parameter.

Figure 9.12 shows $\bar{P}(d_r = 4, d_d) \equiv \bar{P}(d_d)$ as a function of d_d for the system of equations (9.9) with $a = 0.4$ and $b = -0.008$. In this case the transforma-

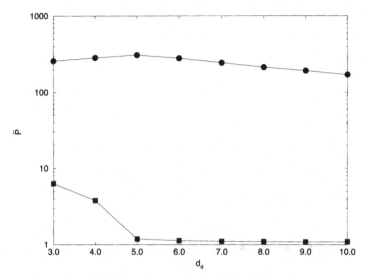

FIGURE 9.12. $\bar{P}(d_d)$ as a function of d_d for different values of the coupling parameter g for time series generated by the system of equations (9.5) and (9.9). The squares represent synchronized behavior at $g = 0.2$, while the circles come from unsynchronized behavior at $g = 0.15$.

tion which maps the drive phase space onto the full phase space is purely algebraic and, therefore, when the systems are synchronized, $\bar{P}(d_d)$ drops to unity as soon as the d_d embedding dimension is high enough to unfold both drive and response time series. When the synchronization manifold loses its stability, $\bar{P}(d_d)$ stays large at all d_d. The square symbols, corresponding to synchronized motion, has $g = 0.2$, while the unsynchronized motions represented by the circles have $g = 0.15$.

In Figure 9.13 we show $(\bar{P}(d_d = 10))^{-1}$ as a function of the coupling coefficient g in Equation (9.9). From the plot alone one can conclude that chaotic motions in the system are synchronized for $g \geq 0.2$ and unsynchronized for $g \leq 0.16$ and that the actual transition to synchronized chaotic behavior occurs somewhere in the interval $0.16 < g < 0.2$. The curve is smeared in this interval as a result of the intermittency discussed above.

9.4   Predictability Tests for Generalized Synchronization

The fact that a functional relationship between the signals from the driving system and the signals from the response system exists, suggests that this function ϕ can be found and used for prediction purposes. Indeed, if two systems are synchronized, then the state of the response system can be predicted solely by knowing the state of a driver, even if the relation

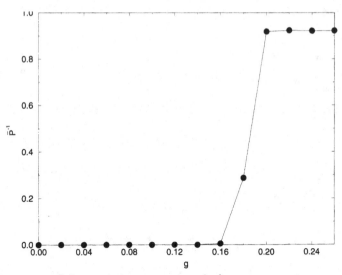

FIGURE 9.13. $\bar{P}^{-1}(d_d = 10)$ as a function of the coupling parameter g in the coupled system Equation (9.5) and Equation (9.9).

between the two variables is rather complicated and unknown. A similar idea was explored in Section 7.4.

We will look for a connection

$$r(n) = G(\mathbf{d}(n)) \tag{9.18}$$

which we can deduce working in the embedding space of the driving system. We assume its dimension to be high enough to unfold the attractor of the driver. Here $\mathbf{d}(n)$ is a vector formed from a scalar time series $d(n)$ in accordance with

$$\mathbf{d}(m) = [d(m), d(m - T_d), d(m - 2T_d), \ldots, d(m - (d_d - 1)T_d)]. \tag{9.19}$$

To find this connection we must have some simultaneous measurements of $d(n)$ and $r(n)$. If such a reconstruction is indeed possible, it signals that the driver and the response are synchronized. If the errors in the prediction become large, that means that there is no synchronization. There are a number of algorithms developed for functional reconstruction as discussed in Chapter 7. Here we again use local polynomial maps. In fact we needed only local linear maps in our work. The nonlinear function $G(\mathbf{x})$ is represented as a collection of local polynomial maps, different for different neighborhoods in the phase space. The parameters of each map are determined by a least-squares fit using known simultaneous measurements of $d(n)$ and $r(n)$.

In order to test the predictive ability of our model we divide our data into two segments. The first segment of N_T pairs $\{d(n), r(n)\}$ we use for

model "training." At the remaining $N_R = N - N_T$ pairs we compare the
predicted values $r_p(n)$ with the measured values $r(n)$ point by point. The
difference

$$(r(n) - r_p(n))/\Delta(r) \tag{9.20}$$

is a useful measure of the quality of prediction. We normalize here by
the standard deviation $\Delta(r)$ of the response signal. It is convenient to
characterize the average quality of prediction by

$$Q = \frac{< (r(n) - r_p(n))^2 >}{2 < (r(n) - < r(n) >)^2 >} \tag{9.21}$$

where

$$< ... > = \frac{1}{N_R} \sum_{i=1}^{N_R} . \tag{9.22}$$

Q should be very small for complete synchronization and will be $O(1)$ if the
quality of prediction is very poor, that is, $r(n)$ and $r_p(n)$ are not correlated.
In Figure 9.14 we present this parameter Q as a function of the coupling parameter g for coupled Rössler systems. In agreement with the
MFNN test, we observe a sharp transition from the synchronized state
to unsynchronized state at the same values of coupling $g \approx 0.18$.

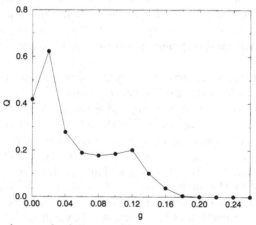

FIGURE 9.14.   Averaged relative prediction error Q computed using Equation
(9.21) for the coupled Rössler systems Equation (9.5) and Equation (9.9) as a
function of the coupling parameter g. Local linear predictors in $d_E = 4$ dimensional
embedding space were built using a learning set of 10,000 pairs $\{d(n), r(n)\}$, and
a time delay $T = 5$ in both cases. Transition to a synchronized state is associated
with a sharp decrease of Q indicating a high degree of predictability. This is where
mutual false nearest neighbors goes to unity as well.

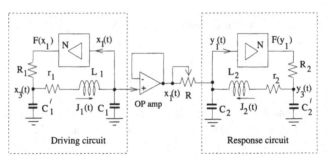

FIGURE 9.15.   The circuit diagram of an experiment with driving and response circuits. Nonlinear converters transform the input X into the output $F(X) = \alpha_i f(X)$. The details of the circuit are discussed in [RVR*92]. An OP amplifier is employed to provide unidirectional coupling. The value of the coupling is controlled by the resistor R. The experimental data was collected by N. F. Rul'kov.

9.4.1   Chaotic Nonlinear Circuits

Now we describe the application of these ideas for detecting generalized synchronization using data from electronic circuits. The data used here was acquired by N. F. Rul'kov using circuits built in his laboratory. The circuit diagram of driving and response circuits coupled by the resistor R is shown in Figure 9.15. The insertion of an operational amplifier into the coupling between the circuits assures that the coupling is one way, namely, from the driving circuit to the response circuit. In the experiment the parameters of the circuits were chosen to be $C_1 = 334nF$, $C_2 = 331nF$, $C_1' = 221nF$, $C_2' = 219nF$, $L_1 = L_2 = 144.9mH$, $R_1 = 4.01k\Omega$, and $\alpha_1 = \alpha_2 = 22.5$. The values of R_2 and R serve as control parameters. Data from driving and response circuit were collected with a sampling rate $50\,\mu s$.

The chaotic attractor generated by the drive circuit is shown in Figure 9.16 as a projection onto the (x_1, x_3) plane of drive system variables. Following our general notation, the drive variables are denoted $\mathbf{x}(t)$ and the response variables are denoted $\mathbf{y}(t)$. When the parameters of the response circuit are tuned to be the same as those of the driving circuit $(R_2 = R_1)$, and the coupling is strong enough, the circuits produce identical synchronized chaotic oscillations. This synchronized behavior can be easily detected by the analysis of projection of the trajectory onto plane of the variables (x_1, y_1) as one can see in Figure 9.17. This plot shows a straight line which is a projection of the manifold of synchronized motions. When the coupling becomes less than some critical value $(R \simeq 0.35\,k\Omega)$ the manifold loses stability and the chaotic oscillations in the response circuit become unsynchronized. The projection onto the (x_1, y_1) plane for $R = 1.0\,k\Omega$ is shown in Figure 9.18 for the unsynchronized oscillations.

It was shown in [RVR*94] that the circuits can display generalized synchronization. To investigate the transition from this generalized synchro-

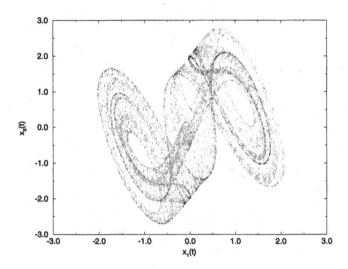

FIGURE 9.16.   The projection of the experimentally observed chaotic attractor of the nonlinear driving circuit seen in Figure 9.15 onto the $(x_1(t), x_3(t))$ plane.

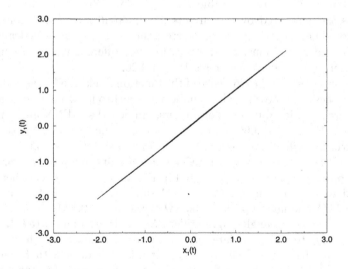

FIGURE 9.17.   The projection of the chaotic attractor onto the $(x_1(t), y_1(t))$ plane observed in the circuits of Figure 9.15 with identical parameters. This comes from synchronized motions where $R = 0.2\,k\Omega$.

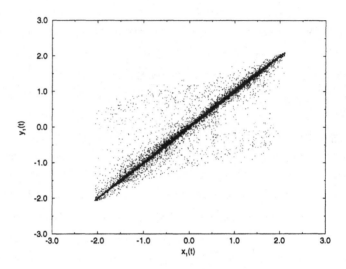

FIGURE 9.18.   The projection of the chaotic attractor onto the $(x_1(t), y_1(t))$ plane observed in the circuits of Figure 9.15 with identical parameters. This comes from unsynchronized motions where $R = 1.0\,k\Omega$.

nization to unsynchronized oscillations, the parameter R_2 was detuned in the response circuit to the value $R_2 = 2.71\,k\Omega$. With this value chaotic motions are generated by the circuits for different values of coupling parameters. The chaotic attractors corresponding to synchronized and unsynchronized oscillations, which occur for two different values of coupling parameter, are shown in Figures 9.19 and 9.20.

In order to test for the existence of the functional relation between driving and response trajectories which indicates synchronized chaotic behavior of the circuits with different parameters, employ the MFNN analysis just described. We used 17,000 data points in each time series for phase space reconstruction which was done with $T_r = T_d = 5$ and $d_r = 5$. Various $\bar{P}(d_d)$ calculated from the data generated by the circuits for different values of the coupling parameter are presented in Figure 9.21. The curves here correspond to $R = 0.0$ circles, $R = 0.1\,k\Omega$–squares, $R = 0.15\,k\Omega$–diamonds, $R = 0.2\,k\Omega$–triangles up, $R = 0.3\,k\Omega$–stars, $R = 0.5\,k\Omega$–triangles left, and $R = 1.0\,k\Omega$—triangles right. When R equals zero and $0.15\,k\Omega$, $\bar{P}(d_d)$ remains close to unity for large d_d, and clearly indicates synchronization between the driving and the response circuits. For data with R equal to $1.0\,k\Omega$ and $0.5\,k\Omega$ large values of $\bar{P}(d_d)$ at high d_d means that the motions in the circuits are unsynchronized.

Figure 9.22 depicts $(\bar{P})^{-1}$ for $d_d = 10$ as a function of the resistance R of the coupling. One can see that this parameter is small for $R = 0.5 - 1.0\,k\Omega$,

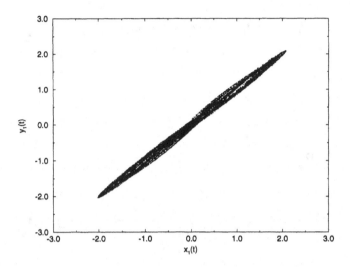

FIGURE 9.19.   The projection of the experimentally observed chaotic attractor of the nonlinear circuit in Figure 9.15 on the $(x_1(t), y_1(t))$ plane when the circuits have different parameters. This comes from synchronized motions where $R = 0.15\,k\Omega$.

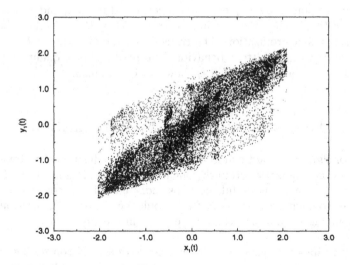

FIGURE 9.20.   The projection of the experimentally observed chaotic attractor of the nonlinear circuit in Figure 9.15 on the $(x_1(t), y_1(t))$ plane when the circuits have different parameters. This comes from unsynchronized motions where $R = 0.5\,k\Omega$.

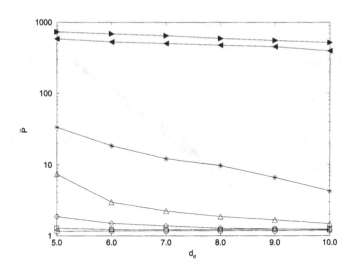

FIGURE 9.21. $\bar{P}(d_d)$ as a function of d_d for the data collected from the coupled electronic circuits: $R = 0.00\,k\Omega$ (circles); $R = 0.10\,k\Omega$ (squares); $R = 0.15\,k\Omega$ (diamonds); $R = 0.20\,k\Omega$ (triangles up); $R = 0.3\,k\Omega$ (stars); $R = 0.50\,k\Omega$ (triangles left); and $R = 1.00\,k\Omega$ (triangles right). Synchronization corresponds to $\bar{P}(d_d)$ near one.

which corresponds to completely unsynchronized motion, and saturates at about 0.6 to 0.7 when the coupling is strong $(0.0 \leq R \leq 0.2\,k\Omega)$ indicating the onset of synchronization. The transition interval $0.2\,k\Omega < R < 0.5\,k\Omega$ corresponds to intermittent behavior. The predictability quality factor Q for this data as a function of R is shown in Figure 9.23.

9.5   A Few Remarks About Synchronization

Synchronization of chaotic motions among coupled dynamical systems is an important generalization from the phenomenon of synchronization of linear systems. The latter is useful, perhaps indispensable, in communications, and one would hope that similar use would be made of the capability of synchronizing in motions which have more interesting structure in phase space.

The two ideas of synchronizing chaotic motions and controlling chaotic motions have common roots in the notion of driving a nonlinear system to restrict its motion on a subspace of the total phase space of the dynamical system. In each case one selects parameter regimes or external forcing to achieve this collapse of the full phase space to a selected subspace. While

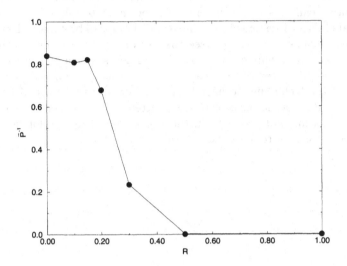

FIGURE 9.22. $\bar{P}^{-1}(d_d = 10$ as a function of the value of the coupling resistor R for the experiment with coupled nonlinear electronic circuits.

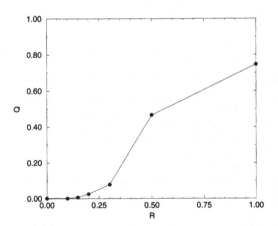

FIGURE 9.23.   Averaged relative prediction error Q computed by formula (9.21) for the coupled nonlinear electronic circuits as a function of resistance R. Local linear predictors in a $d_E = 4$ dimensional embedding space were built using a learning set of 10,000 pairs $\{d(n), r(n)\}$, and a time delay $T = 5$ in both cases. Transition to a synchronized state is associated with a sharp decrease of Q indicating a high degree of predictability. This is where mutual false nearest neighbors goes to unity as well.

this has been successful in many examples, there is room for interesting mathematical analysis of the general circumstances under which this kind of collapse can be stable. That may be too much to ask for, so a sense of general classes of systems in which such collapse can be expected would be equally valuable. Personally I see the use of synchronization, both as an identity among dynamical variables in coupled systems and in the generalized sense discussed in this chapter, as potentially the most commercializable of the techniques to have been uncovered in the study of nonlinear systems with chaotic motions. Careful attention to the theory underlying synchronization and extensive further work toward its development as an engineering design tool are likely to have high payoff.

10
Other Example Systems

In this chapter we discuss in some detail three example systems which have been analyzed using the methods of this book. The examples are quite varied and range over

    i chaotic intensity fluctuations in a solid state laser,

    ii volume fluctuations of the Great Salt Lake in Utah, and

    iii wall pressure fluctuations along the surface of an axisymmetric body rising under its own buoyancy in a deep lake.

In each example we will try to provide some detail about the experiments or observations so one can see the context in which the analysis tools are useful. In the first example we will also present a detailed physical model as an example of how one can use the nonlinear time series analysis as a guide to developing realistic, physically based differential equations for the properties of the source of the observations. In the case of the Great Salt Lake such a detailed model is not yet available. In the case of the wall pressure fluctuations the very high Reynolds number, nearly 100,000 in units based on the boundary layer thickness would seem to indicate that a full blown implementation of the Navier-Stokes equations might be called for. As it happens the sensors in this last example were quite large compared to the finest scales of fluid motion in the boundary layer, so much of the small scale dynamics was filtered out, and the dynamics of the larger spatial scale motions of coherent structures dominated the observations. The dynamical equations for the coherent vortices in the boundary layer are still complicated, but represent a great simplification over the full three-dimensional fluid dynamical equations which, no doubt, are required to describe the details of the fluid motion at these high Reynolds numbers.

In each example we are observing signals which are highly filtered versions of the full dynamical structure. In the example of the chaotic laser, we do not need to worry about the details of the atomic physics in the active medium nor do we worry about the details of the interaction of the lasing light with the intracavity nonlinear crystal nor do we consider the details of the interaction of the light with the mirrors of the laser cavity. Instead a lumped model describing the few modes of lasing light, the frequency doubled light coming from the nonlinear crystal and two nominal atomic levels are used. This kind of modeling is strongly suggested by the analysis of the observed data, and is the kind of physical situation where one can hope to make substantial progress with the methods discussed in this book.

The Great Salt Lake volume data is naturally filtered by the lake itself. It has great spatial extent and measurements were made at regular intervals of two weeks. Details of local rainfall or local stream flow rates or other highly variable quantities were removed from the observations of lake volume in a natural fashion. This allows us to anticipate that only a few degrees of freedom would remain in the data. In other data sets from geophysical sources, including many years of Central Europe pressure data and long observations of internal waves off the coast of California, one sees high dimensional dynamical activity. This arises from the observation of the geophysical details of the processes involved. The methods we have described will not give much insight into such data sets and will not allow much headway in the physical interpretation of those data.

The boundary layer flow data was automatically filtered spatially by the sensors used in the experiment. This filtering removed the finest details of the fluid flow, but serendipitously enhanced the role of coherent structures in the observations. To the extent that the coherent, large scale structures are of interest dynamically, just the right set of observations were made. Were one to have measured the fluid details at the finest scales, the dimension as perceived by false nearest neighbors or other methods would have been large indeed.

10.1   Chaotic Laser Intensity Fluctuations

In the operation of an Nd:YAG laser with an intracavity Potassium Titanyl Phosphate (KTP) crystal, irregular fluctuations of the total output intensity in infrared light ($\lambda \approx 1.064 \, \mu$) are commonly observed. These fluctuations are chaotic as they have at least one positive Lyapunov exponent associated with their evolution [AGLR95, BR91, ABST93], and they have a broad, continuous Fourier power spectrum with a peak near 60 kHz. We are interested in aspects of this chaotic laser which are tied to the quantum mechanical generation of green light ($\lambda \approx 0.532\,\mu$) via the nonlinear susceptibility of the KTP crystal. The green light leaves the cavity as one of the mirrors is transparent at 532 nm.

The laser was set up to operate with three cavity modes of infrared light. One observes two distinctly different irregular time series depending on the polarizations of the light. When the modes are all polarized parallel to each other, the chaotic oscillations of the total infrared intensity shows clear low dimensional behavior as seen via global false nearest neighbors. This class of oscillations, termed Type I chaos, was also accompanied by a very low level of green light. This suggests that the coupling mechanism in the KTP crystal which generates green photons from two infrared photons is very weak when the polarizations of the infrared fields are parallel. In this setup the quantum fluctuations associated with the green light generation were small, and we anticipate that the dynamical equations governing the intensity of the infrared light and the gain in the active medium would be quite accurate in their semi-classical form without quantum fluctuation terms.

A second kind of chaotic motion, Type II chaos, occurs when one of the modes of infrared light is polarized perpendicular to the other two. In this case the production of green light is much stronger, and false nearest neighbors shows that this dimension is not small. The "noise" seen by this statistic is associated with the intrinsic quantum fluctuations accompanying the generation of the green light. Because the apparent noise levels are higher we anticipate that the dynamical equations may be semi-classical but involve significant fluctuation terms.

10.1.1 *Experimental Setup and Data Preparation*

The basic elements of the laser system are a Nd:YAG crystal pumped by a diode laser which is in the same cavity as a KTP crystal. The intracavity nonlinear KTP crystal serves as the frequency doubling element. One end of the laser cavity is formed by the high reflection coated flat facet of the Nd:YAG crystal. This facet is highly reflecting at both the infrared fundamental (1064 nm) and at the doubled green (532 nm) wavelength and is highly transmissive at the pump wavelength (810 nm). The back end of the laser cavity is formed by a curved output coupler which is highly transmitting for the green light and highly reflecting for infrared light [BR91]. The green light is thus not resonant in the cavity and acts as a loss mechanism.

The laser system can display steady state, periodic, quasi-periodic and chaotic intensity fluctuations when operated with three or more longitudinal modes. Without the KTP crystal in the cavity these oscillations are normally heavily damped and stabilized by this damping. The doubling process provides the nonlinear mechanism which destabilizes these oscillations. The total intensity in the infrared $i(t)$ was observed with a photodiode having a rise time of less than 1 ns and was sampled using a 100 MHz eight bit digital oscilloscope capable of storing 10^6 samples.

In the time traces we can see the distinction between these two operating regimes. Type I consists of long "bursts" of relaxation oscillations, while Type II appears far more irregular. During Type I operation very little green light, less than 1 μW, was observed, while more than 25 μW of power in green light accompanied Type II activity. This is consistent with the linear stability properties of the macroscopic equations describing this laser [BR91]. If all three modes are parallel polarized as in Type I behavior, the laser can become unstable with a very small coupling in the KTP crystal, but very little green light is produced. If one of the modes is polarized perpendicular to the other two, a very small value coupling then results in appreciable sum frequency generation, instability to chaotic operation, and roughly two orders of magnitude larger intensity of green light than in Type I operation.

The resolution of the digital sampling oscilloscope at eight bits or 0.4 percent is too low for some of the data analysis, notably the Lyapunov exponent calculation. In order to improve the resolution of the data sets and still have the large number of points need to analyze data in high dimensions, the oscilloscope was set to sample the data at $f_s = 10$ MHz and the entire 10^6 sample memory was used. This rate is 200 times higher than the 60 kHz relaxation oscillations of the dynamics ensuring that the detailed evolution of the signal was captured and aliasing did not occur. For the calculations where resolution was not critical, the data sets were downsampled by a factor of resulting in an effective sampling rate of 1.25 MHz or a sampling period of 800 ns. This rate is still 20 times larger than the relaxation oscillation rate, and as a result the first minimum of the mutual information is found at four or five samples or one quarter of the relaxation oscillation period. Note that in downsampling, the broadband noise level due to high-dimensional dynamics is neither increased or decreased. When the eight bit resolution was insufficient, the raw data sets were interpolated using a digital linear filter. This filter was designed to remove frequencies from 500 kHz to the Nyquist frequency $f_s/2 = 5$ MHz and pass all frequencies below 500 kHz. Since the signal due to the dynamics alone had no frequency as high as 500 kHz, no dynamical information was lost. On the other hand, the quantization noise, which is assumed to be white up to the Nyquist frequency, was cut down by 90 percent. Thus, by reducing the quantization noise or error by a factor of ten, we gained slightly more than three bits of resolution. After performing the digital convolution, the data was downsampled from the original 10 MHz to 1.25 MHz. Since the filter had already removed frequencies above 500 kHz, no aliasing occurred and the data had 11 bits of resolution.

Laser Output Intensity

Type I Chaos

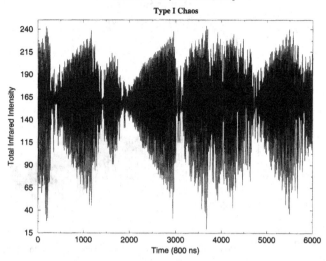

FIGURE 10.1.   A typical time trace for Type I chaos. The sampling time is $\tau_s = 800$ ns. Six thousand points out of the 125,000 observed are displayed. The laser was operating with three infrared cavity modes all polarized parallel to each other.

10.1.2   Analysis of the Chaotic Laser Data

For data of Type I chaos, we display in Figure 10.1 a sample time series from our observations. The sampling time is $\tau_s = 800$ ns. Figure 10.1 shows 6,000 data points out of the 125,000 collected. The Fourier power spectrum of these data is in Figure 10.2, and we can see broad spectral features near 60 kHz and apparent harmonics of that frequency. The broadband nature of the spectrum is characteristic of chaotic motion. Figure 10.3 shows the average mutual information evaluated from these data. We see that the average mutual information has its first minimum near $T\tau_s = 5\tau_s = 4\,\mu$ sec. This was the location of the first minimum in $I(T)$ for each of the Type I data sets we examined. For data from a Type II chaos trace, the time series is shown in Figure 10.4 and the Fourier spectrum in Figure 10.5. $\tau_s = 800$ ns again. The spectral features seen in Type I chaos are washed out with other spectral "peaks" visible in these data. Once again we evaluated the average mutual information and see a typical result in Figure 10.6 from one of our data sets we examined. The first minimum for $I(T)$ is again near $T = 5$ for each of the Type II data sets. Here it was at $T = 4$. For both a Type I data set and a Type II data set the percentage of false nearest neighbors is shown in Figure 10.7 for an average over a series of Type I and of Type II measurements. The interpretation of this is quite clear: the attractor associated with Type I chaos is captured in a low dimensional space with

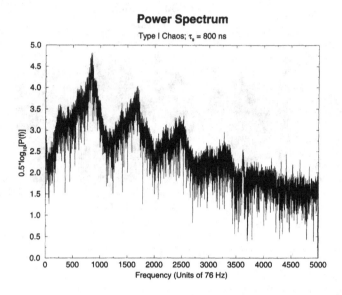

FIGURE 10.2.   The Fourier Power Spectrum of the Type I chaos seen in Figure 10.1. There is a broad peak in the spectrum near 60 kHz along with other broad spectral features.

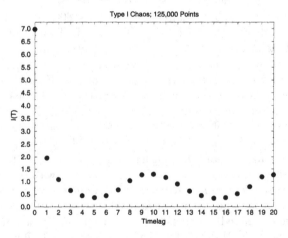

FIGURE 10.3.   The average mutual information $I(T)$ for Type I chaos. One hundred twenty five thousand samples were used. A clear first minimum at $T = 5$ corresponding to 4μ sec is seen.

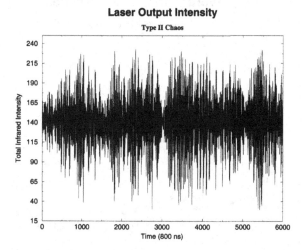

FIGURE 10.4.   A typical time trace for Type II chaos. The sampling time is $\tau_s = 800$ ns. Six thousand points out of the 125,000 observed are displayed. The laser was operating with three infrared cavity modes where two are polarized parallel to each other and the other is polarized perpendicular to the first two.

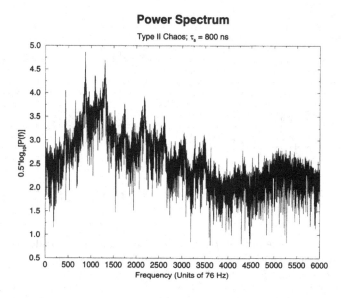

FIGURE 10.5.   The Fourier Power Spectrum of the Type II chaos seen in Figure 10.4. The broad peak near 60 kHz seen in Type I chaos, Figure 10.2, has been further broadened.

Average Mutual Information

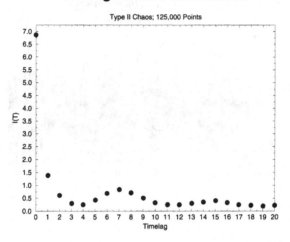

FIGURE 10.6.   The average mutual information $I(T)$ for Type II chaos. 125,000 samples were used. A clear first minimum at $T = 4$ corresponding to 3.2μ sec is seen.

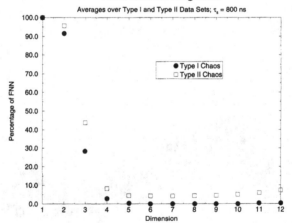

FIGURE 10.7.   The percentage of global false nearest neighbors for both Type I (solid circles) and Type II chaos (open squares). These are averages over a collection of Type I and Type II data. For Type I chaos we used $T = 5$ or $T = 6$ for the state space reconstruction while for Type II, $T = 3$ or $T = 4$. Both operating conditions show some quantum noise from the production of green light, but the levels of residual false neighbors is much higher in Type II chaos where the green production is substantially larger. One hundred twenty five thousand samples were used in each case, and $\tau_s = 800$ ns.

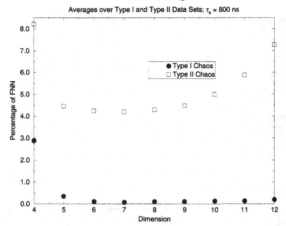

False Nearest Neighbors

Averages over Type I and Type II Data Sets; τ_s = 800 ns

FIGURE 10.8. An enlargement of the global false neighbors statistic seen in Figure 10.7 to emphasize the residual level of false neighbors for Type II chaos. The Type II level is nearly 40 times that for Type I.

$d_E \approx 5$. Our analysis of Type II chaos yields quite a different result. We show in Figure 10.8 the false nearest neighbors for $4 \leq d_E \leq 10$ for this average over data sets of Type I and Type II. We see that the percentage of global false nearest neighbors for Type I chaos falls to zero at $d_E = 7$. The percentage of false nearest neighbors does not fall to zero for any dimension for the Type II data, and this indicates the presence of a high dimensional signal in these data. The high residual level of global false nearest neighbors is about 40 or more times higher for Type II data. We interpret this level of high dimensional signal as noise arising from the intrinsic quantum fluctuations in the process which gives rise to Type II chaos. In particular it is observed that when the chaos is Type II, the infrared light is accompanied by a large amount of green light associated with the conversion of infrared photons in the KTP crystal. In Figure 10.9 we show the fraction of bad predictions P_K for Type I chaos when we predict forward in time. Clearly P_K becomes independent of the local dimension d and of N_B at $d_L = 7$. If we predict backward in time, as shown in Figure 10.10, we draw the same conclusion. This is consistent with the behavior of Type I chaos seen in Figure 10.8 where the number of global false nearest neighbors falls to zero at $d_E = 7$, then remains there. Local false nearest neighbors are much more sensitive to fine structure on the attractor than global false neighbors. The latter presents a kind of global average over all regions of phase space, so regions which exhibit the highest dimensional structure may occupy only a small percentage of the total phase space. This

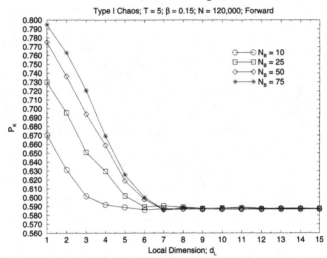

FIGURE 10.9.   Local false nearest neighbors for Type I chaos using $T = 5$ and 120,000 points from the time series. The computation was done forward in time. There is a clear indication that at $d_E = 7$ the predictability P_K of these data has become independent of the number of neighbors and the embedding dimension. β, here set to 0.15, defines the size of the error ball within which a good prediction must fall after T steps forward in predicting. The error ball is β times the overall size of the attractor. See [AK93] for more details on these parameter choices in this algorithm.

small percentage of space regains proper importance when local quantities are computed, as we are doing now.

In Figures 10.11 and 10.12 we show the result of the calculation of forward then backward local false nearest neighbors for Type II chaos. Note that the P_K values are much higher for Type II chaos reflecting the presence of high dimensional quantum noise in the data. This noise is not of such high amplitude as to ruin completely the possibility of making local predictions, but it certainly enormously erodes the quality of those predictions. The level of bad predictions for the same parameter settings in the false nearest neighbor algorithm for both types of chaos leads to about 20% bad predictions for Type I chaos and nearly 60% bad predictions for Type II chaos. At the same time there is a clear indication that the dynamical dimension of the system giving rise to the observations is $d_L = 7$ in each case. This is a very nice result in that it shows that low levels of noise do not impede our ability to identify the number of differential equations required to describe the data.

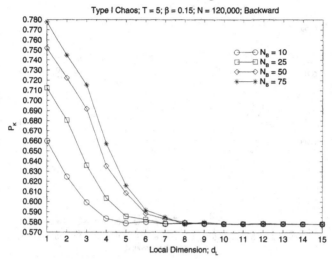

FIGURE 10.10.   Local false nearest neighbors for Type I chaos using $T = 5$ and 120,000 points from the time series. The computation was done backward in time. There is a clear indication that at $d_E = 7$ the predictability P_K of these data has become independent of the number of neighbors and the embedding dimension. β, here set to 0.15, defines the size of the error ball within which a good prediction must fall after T steps backward in predicting. The error ball is β times the overall size of the attractor. See [AK93] for more details on these parameter choices in this algorithm.

Seven degrees of freedom could well have been anticipated from physical reasoning. We have three modes of the infrared field each of which has an electric and a magnetic field describing it. So we have six degrees of freedom from the electromagnetic field. The green field is significantly damped by its not being a cavity mode, so associated with that field we should expect some large and negative Lyapunov exponents. Finally we have the atomic degrees of freedom in the active medium. We anticipate a single gain equation associated with the population inversion of the level responsible for the principal transition near $1.064\,\mu$. This will give us the seven degrees of freedom seen in the experimental data.

If we provide more degrees of freedom in the form of differential equations for Heisenberg operators or density matrix elements, we expect to find large, negative Lyapunov exponents associated with the damping of these quantities. The local false nearest neighbors results tells us the number of "active" degrees of freedom out of the many we could anticipate entering this problem. "Active degrees of freedom" is best defined by the example we are discussing now, namely, those dynamical variables which

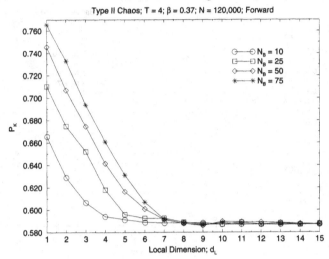

Local False Neighbors

Type II Chaos; T = 4; β = 0.37; N = 120,000; Forward

FIGURE 10.11.   Local false nearest neighbors for Type II chaos using $T = 4$ and 120,000 points from the time series. The computation was done forward in time. There is a clear indication that at $d_E = 7$ the predictability of these data has become independent of the number of neighbors and the embedding dimension. Note the much higher percentage of **unpredictable** points on the attractor here compared to Type I chaos. The level of unpredictability for $d \geq d_L$ is nearly three times that seen for Type I chaos. This is a direct result of the higher quantum noise level in Type II chaos. Nonetheless the local false nearest neighbor statistic is seen to be quite robust against noise. β, here set to 0.37, defines the size of the error ball within which a good prediction must fall after T steps forward in predicting. The error ball is β times the overall size of the attractor. See [AK93] for more details on these parameter choices in this algorithm.

are not substantially removed from the dynamical description of the physical situation by damping or losses. Once again those variables which are suppressed by the losses would show up in exhibiting large, but negative, Lyapunov exponents were we to evaluate them in dimensions larger than that indicated by local false neighbors, here $d_L = 7$.

Using $d_E = d_L = 7$, we have computed the $\lambda_a(\mathbf{x}, L)$ for a large number of starting locations $\mathbf{x}$ on the attractor, then determined the value of these quantities as a function of the number of steps we look ahead of these starting points. We used 5000 starting points and carried the calculation out 2048 steps ahead of each of these locations. This allows us to evaluate the average local Lyapunov exponents. These are shown in Figure 10.13 and then in an enlarged view, namely, for $7 \leq L \leq 11$ in Figure 10.14 for Type I chaos. There are clearly two positive Lyapunov exponents, one zero exponent which is characteristic of the dynamics of differential equations [ABST93], and four negative exponents. Their sum is negative, as it

Local False Neighbors

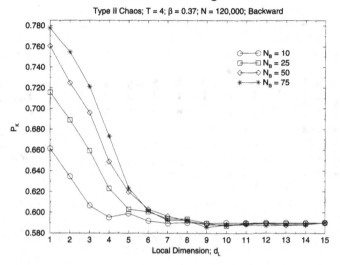

FIGURE 10.12.   Local false nearest neighbors for Type II chaos using $T = 4$ and 120,000 points from the time series. The computation was done backward in time. There is a clear indication that at $d_E = 7$ the predictability of these data has become independent of the number of neighbors and the embedding dimension. Note the much higher percentage of **unpredictable** points on the attractor here compared to Type I chaos. The level of unpredictability for $d \geq d_L$ is nearly three times that seen for Type I chaos. This is a direct result of the higher quantum noise level in Type II chaos. Nonetheless the local false nearest neighbor statistic is seen to be quite robust against noise. β, here set to 0.37, defines the size of the error ball within which a good prediction must fall after T steps backward in predicting. The error ball is β times the overall size of the attractor. See [AK93] for more details on these parameter choices in this algorithm.

should be, and for this sample of Type I chaos that sum is approximately -1.28 in units of inverse τ_s. Each of the average local exponents shown here changes sign when the eigenvalues of the Oseledec matrix is evaluated backward in time. This is as it should be for real exponents, and this supports our choice of $d_E = d_L = 7$ from local false neighbors.

The same calculations were done on our data set for Type II chaos. From the outset we must recall that when a data set is noisy, the evaluation of Lyapunov exponents may well be uncertain [ABST93]. The origin of the uncertainty is in the severely ill-conditioned nature of the Oseledec matrix which serves to amplify any numerical errors in the individual Jacobians composing it. With noise the determination of neighboring distances and the local map from which we read off the Jacobian is sure to lead to small errors in each Jacobian. This will lead to real uncertainties in the Lyapunov exponents, especially the negative exponents [ABST93].

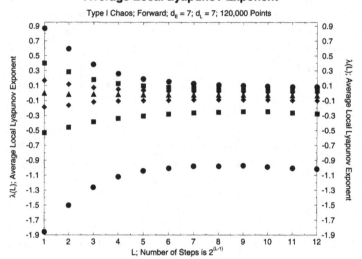

FIGURE 10.13.   The average local Lyapunov exponents for our sample of Type I chaos. A global and a local dimension of $d_E = d_L = 7$ has been used, and 120,000 points from the time series were utilized. There are two positive Lyapunov exponents and one zero exponent indicating that differential equations describe the underlying dynamics. The Lyapunov dimension D_L for this is about 4.95 ± 0.1 telling us that the last large, negative exponent is not very important dynamically. This is consistent with the fall of the global false nearest neighbors, shown in Figure 10.7, to nearly zero by dimension 5. The attractor, which is essentially five dimensional, is twisted significantly in the time delay coordinate system provided by the vectors $\mathbf{y}(n)$, and it requires seven dimensions to completely unfold its intersections with itself.

Nonetheless, we have evaluated the average local Lyapunov exponents for these data, and in Figure 10.15 we present the results of this calculation. Figure 10.16 shows an enlargement of the average exponents for large numbers of steps along the trajectory after a perturbation to the known orbit. From these figures we see that the largest exponent is about twice that of the largest Type I exponent. This would lead immediately to increased unpredictability, as we saw in the evaluation of local false nearest neighbors. Also we see three positive exponents, which would be connected to the apparent inability to control the Type II chaos with the methods which have been tried [RJM*92, GIR*92]. It is reassuring that one of the exponents is zero, so we again have a set of differential equations describing the source of these data.

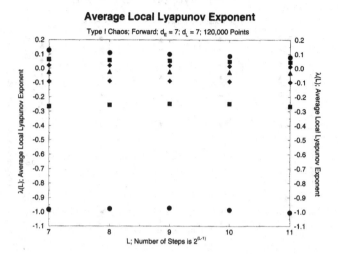

FIGURE 10.14.   A blowup of the average local Lyapunov exponents shown in
Figure 10.13. The zero exponent and the two positive exponents are much clearer
in this view of the data.

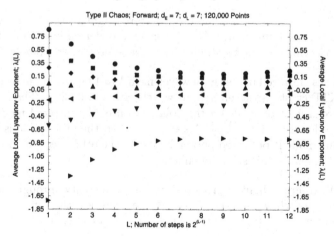

FIGURE 10.15.   The average local Lyapunov exponents for our sample of Type
II chaos. A global and a local dimension of $d_E = d_L = 7$ has been used, and
120,000 points from the time series were utilized. There are three positive Lyapunov
exponents and one zero exponent indicating that differential equations describe the
underlying dynamics.

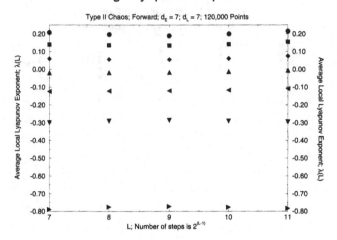

FIGURE 10.16.   A blowup of the average local Lyapunov exponents shown in Figure 10.15. The zero exponent and the three positive exponents are much clearer in this view of the data.

10.1.3   A Physical Model for the Process

The salient features of the data analysis are

i   There are only a few modes of infrared light in the laser cavity. Indeed, the number in this experiment was determined to be three by observing the infrared light in a Fabry-Perot interferometer.

ii These infrared modes with wavelength $\lambda \approx 1.064\,\mu$ couple through the KTP crystal to green light at wavelength $\lambda \approx 0.532\,\mu$. The amount of green light depends on the polarization of the infrared modes. If all modes are polarized parallel to each other, the production is small. If one mode is polarized perpendicular to the other two, the production of green light is strongly enhanced.

iii When green light is produced, it exits the cavity through one of the mirrors which is transmitting at that wavelength. The same mirror reflects the infrared light, so green is not a cavity mode.

We neglect the detailed dynamics of the active medium, treating it as a two level quantum system. Nd:YAG is more complicated but no changes in our essential conclusions will come from these complications. In the two level approximation, the atomic systems have an upper level $|u>$ and a lower level $|l>$ with energy difference $\hbar\omega_A$. The dynamical variables for the atomic levels are the usual Pauli spin operators. We have annihilation and creation operators $a^\dagger_m(t)$ and $a_m(t)$ for the M infrared modes satisfying

$$[a_m, a^\dagger_n] = \delta_{mn}. \tag{10.1}$$

The indices m or n refer both to the mode and to the polarization. Green light is treated as a scalar field with creation and annihilation operators $g^\dagger(t)$ and $g(t)$ satisfying

$$[g, g^\dagger] = 1, \tag{10.2}$$

at equal times. Of course, the green light has polarization and probably more than a single mode, but these were not measured. The green modes are damped out so strongly relative to the infrared that they trail or are fully determined by the infrared dynamics. The green modes play a "non-dynamical" role in these experiments. The loss of green light through the mirror will be treated in a conventional manner [Kim92] as a coupling to a "reservoir" of radiation modes $r^\dagger_k$ and r_k which are also bosons. The loss mechanisms for the atomic levels, essentially the electromagnetic modes responsible for spontaneous emission from the upper level $|u>$, are labeled R_a and $R^\dagger_a$. These modes of the electromagnetic field serve solely to describe the loss mechanism as seen in the lasing system.

The Hamiltonian for the system is written as

$$
\begin{aligned}
H = {}& \frac{\hbar\omega_A}{2}S_3 + \sum_{l=1}^{M}\hbar\omega_l a^\dagger_l a_l + \hbar\omega_g g^\dagger g + \sum_k \hbar\Omega_k r^\dagger_k r_k + \sum_a \hbar\omega_a R^\dagger_a R_a \\
& +i\hbar g^\dagger\left[\sum_{l,m=1}^{M}\kappa_{lm}a_l a_m\right] - i\hbar\left[\sum_{l,m=1}^{M}\kappa^*_{lm}a^\dagger_l a^\dagger_m\right]g \\
& +i\hbar g^\dagger\sum_k\Gamma_k r_k - i\hbar\left[\sum_k\Gamma^*_k r^\dagger_k\right]g \\
& +i\hbar S_+\left\{\sum_{m=1}^{M}\chi_m a_m + \sum_a\gamma_a R_a + Ee^{-i\omega_p t}\right\} \\
& -i\hbar\left\{\sum_{m=1}^{M}\chi^*_m a^\dagger_m + \sum_a\gamma^*_a R^\dagger_a + E^*e^{i\omega_p t}\right\}S_-
\end{aligned}
\tag{10.3}
$$

In this expression, the ω_l are the frequencies associated with the infrared modes; ω_g is the green frequency: $\omega_g \approx 2\omega_l$. Ω_k are the frequencies of the reservoir modes coupled to the green light, and ω_a are the reservoir frequencies for the atomic loss mechanism. The nonlinear coupling κ_{lm} comes from the second order susceptibility of the KTP crystal. It is dependent on the polarizations of the infrared photons. The Γ_k tell us the strength of the coupling of the green modes to the reservoir, and γ_a is the coupling of the atomic levels to the reservoir modes. Finally E represents the strength of the external pumping at frequency ω_p in its coupling to the atomic levels.

There is no damping of the infrared modes included in this model Hamiltonian though there is no reason in principle why these modes are not also damped. The main effect we observe is tied to the relatively high transparency of the cavity mirrors to the green light, so any damping of the infrared modes is unlikely to play a major role here.

Using this Hamiltonian and the standard Heisenberg equations of motion

$$i\hbar \frac{d\bullet}{dt} = [\bullet, H], \tag{10.4}$$

we arrive at

$$\frac{da_n(t)}{dt} = -i\omega_n a_n(t) - \sum_{m=1}^{M} a^\dagger{}_m (\kappa^*_{nm} + \kappa^*_{mn}) g(t) - \chi^*_n S_-(t),$$

$$\frac{dg(t)}{dt} = -i\omega_g g(t) + \sum_{l,m=1}^{M} \kappa_{lm} a_l(t) a_m(t) + \sum_k \Gamma_k r_k(t),$$

$$\frac{dr_k(t)}{dt} = -i\Omega_k r_k(t) - \Gamma^*_k g(t),$$

$$\frac{dR_a(t)}{dt} = -i\omega_a R_a - \gamma^*_a S_-(t),$$

$$\frac{dS_3(t)}{dt} = 2S_+(t)\left\{ \sum_{m=1}^{M} \chi_m a_m(t) + \sum_a \gamma_a R_a(t) + E e^{-i\omega_p t} \right\},$$

$$+2\left\{ \sum_{m=1}^{M} \chi^*_m a^\dagger{}_m(t) + \sum_a \gamma^*_a R^\dagger{}_a(t) + E^* e^{i\omega_p t} \right\} S_-(t),$$

$$\frac{dS_+(t)}{dt} = i\omega_A S_+(t) - \left\{ \sum_{m=1}^{M} \chi^*_m a^\dagger{}_m(t) \right.$$

$$\left. + \sum_a \gamma^*_a R^\dagger{}_a(t) + E^* e^{i\omega_p t} \right\} S_3(t). \tag{10.5}$$

To eliminate the reservoir coordinates $r_k(t)$, integrate the linear equation for $r_k(t)$:

$$r_k(t) = e^{-i\Omega_k t} r_k(0) - \Gamma^*_k \int_0^t d\tau\, e^{-i\Omega_k(t-\tau)} g(\tau), \tag{10.6}$$

which used in the equation of motion for $g(t)$ gives us

$$\frac{dg(t)}{dt} = -i\omega_g g(t) + \sum_{l,m=1}^{M} \kappa_{lm} a_l(t) a_m(t)$$

$$+ \sum_k \Gamma_k e^{-i\Omega_k t} r_k(0)$$

$$- \int_0^t d\tau \left[\sum_k |\Gamma_k|^2 e^{-i\Omega_k(t-\tau)} \right] g(\tau). \tag{10.7}$$

Now if we make the usual approximation of a continuum in the sum over reservoir modes, then we may replace the integral in the equation of motion for $g(t)$ by

$$\int_0^t d\tau \left[\sum_k |\Gamma_k|^2 e^{-i\Omega_k(t-\tau)} \right] g(\tau) \rightarrow [\Gamma + i\Delta] g(t). \qquad (10.8)$$

Here $\Gamma > 0$ is a decay parameter representing the leakage of green light out to the reservoir, and Δ is an energy level shift due to the interaction of green light with reservoir modes. This approximation is discussed at some length in the Les Houches lectures of Kimble [Kim92]. With this approximation, we have

$$\frac{dg(t)}{dt} = -[\Gamma + i\omega_g']g(t) + \sum_{l,m=1}^M \kappa_{l,m} a_l(t) a_m(t) + \sum_k \Gamma_k e^{-i\Omega_k t} r_k(0), \qquad (10.9)$$

with $\omega_g' = \omega_g + \Delta$.

Since this equation is linear in $g(t)$, we can integrate to find

$$\begin{aligned} g(t) \;=\;& e^{-(\Gamma + i\omega_g')t} g(0) \\ &+ \sum_k \Gamma_k r_k(0) \frac{e^{-i\Omega_k t} - e^{-(\Gamma + i\omega_g')t}}{\Gamma + i(\omega_g' - \Omega_k)} \\ &+ \int_0^t d\tau \, e^{-(\Gamma + i\omega_g')(t-\tau)} \sum_{l,m=1}^M \kappa_{lm} a_l(\tau) a_m(\tau). \end{aligned} \qquad (10.10)$$

In the integral we replace the rapidly varying infrared operators $a_m(t)$ by the more slowly varying interaction representation forms $A_n(t)$ where

$$a_n(t) = e^{-i\omega_n t} A_n(t), \qquad (10.11)$$

then perform the integrations by removing the slowly varying operators from under the integral.

For times large compared to Γ^{-1} the damping in the green field becomes important and the behavior of $g(t)$ is

$$g(t) \approx \sum_k \frac{e^{-i\Omega_k t}}{\Gamma + i(\omega_g' - \Omega_k)} \Gamma_k r_k(0) + \sum_{l,m=1}^M \frac{1}{\Gamma + i(\omega_g' - \omega_m - \omega_l)} \kappa_{lm} a_l(t) a_m(t).$$
$$(10.12)$$

This says the residual green light, after damping becomes effective, is that produced from the quadratic term in infrared annihilation operators along with the associated fluctuations coming from the reservoir modes. The green mode is seen here to be "slaved" to the infrared dynamics, namely, $g(t)$ is determined solely in terms of the infrared modes and fluctuations associated with its coupling to the external world. If we had kept the details of the green light dynamics, it would have led us to the same essential conclusion about the number of active degrees of freedom since each green mode would have been "frozen" out in this way.

Eliminating the $g(t)$ dynamics from the $a_n(t)$ equations we see that the infrared amplitude fluctuations are multiplicative in the $a_n(t)$, not additive. Further, the strength of the fluctuations depends on the couplings κ_{lm} which govern the production rate of green light from infrared light. This provides us with the essential connection between the substantially augmented quantum noise fluctuations seen in Type II chaos and the simultaneous observation in that type of chaos of large amounts of green light. When there is little green light, as in Type I chaos, there is also little quantum fluctuation in the infrared intensities. The latter is what we learned in examining the global false nearest neighbors of the two types of chaotic data.

Next we eliminate the atomic reservoir modes $R_a(t)$ and find that just as the green modes discussed above are determined entirely by the infrared modes, the $S_{\pm}(t)$ are determined by the $S_3(t)$, the infrared modes, the pump, and the atomic reservoir fluctuations. This means that we will be left with equations for the $a_m(t)$, the $a^{\dagger}{}_m(t)$, and $S_3(t)$. Of course these are nonlinear operator equations, so we have not gotten very close to solving the equations, but we have eliminated much of the potential complication in this problem.

We now have reduced the number of degrees of freedom to the smallest number we can expect in this system. From our knowledge of the experimental observation of three infrared modes using a Fabry-Perot interferometer, we can conclude that altogether we require seven equations, nonlinear operator equations to be sure, to describe the laser system at hand. This is just the number selected by our analysis above, and it gives substantial support to that analysis. Further, it suggests that using a semiclassical version of these operator equations achieved by simply replacing the remaining operators by c-numbers and parameterizing the fluctuations using a choice for the reservoir statistics, we should be able to quite nicely describe the experiments in quantitative detail. This would include reproducing the nonlinear statistics we used in our analysis above, especially the local and global Lyapunov exponents. This has been a detailed and explicit example of how we can use the analysis tools which form the central theme of this book to determine the structure of dynamical equations describing the mechanisms of the data source. In this case, as we suspect it must be in any case, we were required to use knowledge of the experiments beyond the ASCII total infrared intensity $i(t)$ with which we began our analysis. We required some precise information about the laser cavity and about the quantum mechanics of light coupling with atomic levels.

10.1.4 Predicting on the Attractor

Even without knowledge of the dynamical equations for the laser system we can use the information we have acquired so far to make predictive models for the laser intensity evolution. The method as described earlier in some generality utilizes the compactness of the attractor in $\mathbf{y}(n)$ space by noting that we have knowledge of the evolution of whole phase space neighborhoods into whole neighborhoods later in time. Here we use only local linear models as we have large amounts of data and thus good coverage of the attractor, namely, every neighborhood is rather well populated.

Recall the idea that locally on the attractor we find the N_B neighbors $\mathbf{y}^{(r)}(n)$; $r = 1, 2, \dots N_B$ of each point $\mathbf{y}(n)$ and make the local linear model $\mathbf{y}(r; n+1) = \mathbf{A}_n + \mathbf{B}_n \cdot \mathbf{y}^{(r)}(n)$ where $\mathbf{y}(r; n+1)$ is the point to which $\mathbf{y}^{(r)}(n)$ goes in one time step. The coefficients $\mathbf{A}_n$ and $\mathbf{B}_n$ are determined by minimizing at each time location n

$$\sum_{r=0}^{N_B} |\mathbf{y}(r; n+1) - \mathbf{A}_n - \mathbf{B}_n \cdot \mathbf{y}^{(r)}(n)|^2, \tag{10.13}$$

where $r = 0$ means $\mathbf{y}(n)$ or $\mathbf{y}(n+1)$. When we have a new point $\mathbf{z}(k)$ on or near the attractor, we seek the nearest neighbor $\mathbf{y}(Q)$ among all the data in the set we used to determine the $\mathbf{A}_n$ and the $\mathbf{B}_n$. The predicted point $\mathbf{z}(k+1)$ is then

$$\mathbf{z}(k+1) \approx \mathbf{A}_Q + \mathbf{B}_Q \cdot \mathbf{z}(k). \tag{10.14}$$

This works remarkably accurately within the limits of prediction dictated by the largest Lyapunov exponent λ_1. When we try to predict beyond the instability horizon, that is for times much greater than

$$\frac{\tau_s}{\lambda_1}, \tag{10.15}$$

our predictions should rapidly lose accuracy.

In Figure 10.17 we show an example of this prediction technique for Type I chaos based on a total data set of 60,000 points. Forty thousand points were used to determine the local polynomial coefficients $\mathbf{A}_n$ and $\mathbf{B}_n$, then predictions were made 10 steps ahead from the point $n = 45,000$ to the point $n = 55,000$. The results for points $n = 45,000$ to $n = 45,500$ are shown in the figure. The predictions are shown as solid circles while the observed data are shown as the solid line. In units of τ_s the largest Lyapunov exponent is approximately $\lambda_1 \approx 0.08$, so we should be able to make accurate predictions out to twelve or so steps beyond any starting location on the attractor. The computations reported in Figure 10.17 support this quite well. In Figure 10.18 we show the result of the same calculation but now predicting ahead 50 steps. There are clearly regions where the predictability is rather good, but also regions where the method starts to fail quite visibly. Figure 10.19 shows a region where the predictability remains quite good,

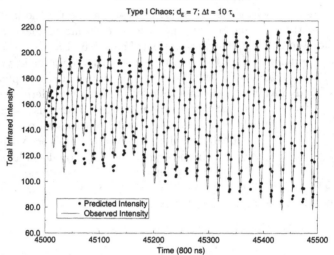

FIGURE 10.17.   Predicted and observed total infrared intensity for Type I chaos. The predictions are made using local linear maps in the reconstructed phase space of the attractor. The coefficients for the local maps are learned from the first 40,000 points of the data set, and then predictions are made starting with point 45,000. All predictions are made in $d_E = d_L = 7$. This figure shows the result of predicting ahead $\Delta t = 10\tau_s$. The largest global Lyapunov exponent is about 1/12 in units of τ_s, so we do not expect to be able to accurately predict ahead much more than this amount.

while Figure 10.20 is a region where predictability is quite reduced for this large time step ahead of a known point. These results are consistent with the wide variation of **local** Lyapunov exponents on an attractor [ABST93]. Pushing this even further we show in Figure 10.21 the result of trying to predict ahead 100 steps beyond a new point near the attractor. Clearly predictability has been substantially lost, as it should be reduced in a chaotic system.

We have also used this method for predicting on the attractor in Type II chaotic data. In Figure 10.22 we display the result of learning on 40,000 points of this data set and then predicting ahead from point 45,000. The predictions ten steps ahead compared to the observed values for $I(t)$ are shown in this figure for time steps 55,000 through 57,500. In Figure 10.23 we have enlarged the region between steps 55,000 and 55,500 so one may see that the quality of these predictions is not as good as we saw in working with Type I data. This is as it should be. Figure 10.24 is another enlargement of the Type II predictions making much the same point about the quality of the predictions. Finally in Figure 10.25 we have predictions 50

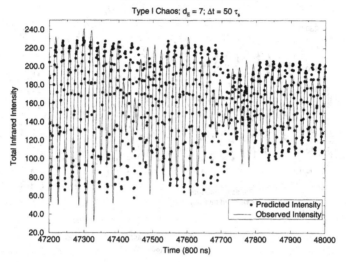

Laser Chaos: Prediction

FIGURE 10.18.   Predicted and observed total infrared intensity for Type I chaos. The predictions are made using local linear maps in the reconstructed phase space of the attractor. The coefficients for the local maps are learned from the first 40,000 points of the data set, and then predictions are made starting with point 45,000. All predictions are made in $d_E = d_L = 7$. This figure shows the result of predicting ahead $\Delta t = 50\tau_s$. The predictions are much worse than for $\Delta t = 10\tau_s$ since we are trying to predict beyond $\frac{\tau_s}{\lambda_1}$.

time steps ahead for Type II chaos. Again 40,000 points were used for learning the local linear maps used for prediction. The quality of the predictions here has become quite poor.

10.1.5   A Few Remarks About Chaotic Laser Fluctuations

This has been quite a detailed look at the chaotic fluctuations of a solid state laser. It illustrates the use of essentially all the tools in our toolkit except signal separation, and with the material in hand, we could have done that as well. The analysis of the chaotic laser fluctuations is special in some important sense. To my knowledge it is the only example to date which starts with scalar observations, namely, the total infrared intensity $i(t)$, and succeeds, by adding knowledge of the dynamical equations for the laser system, in going all the way to equations of motion whose output can be compared to the results of using the toolkit to analyze the data. The comparison would, of course, be made in terms of the statistical quantities which are evaluated by the tools and not in terms of specific time traces for the infrared intensity. Indeed, having equations of motion we are able

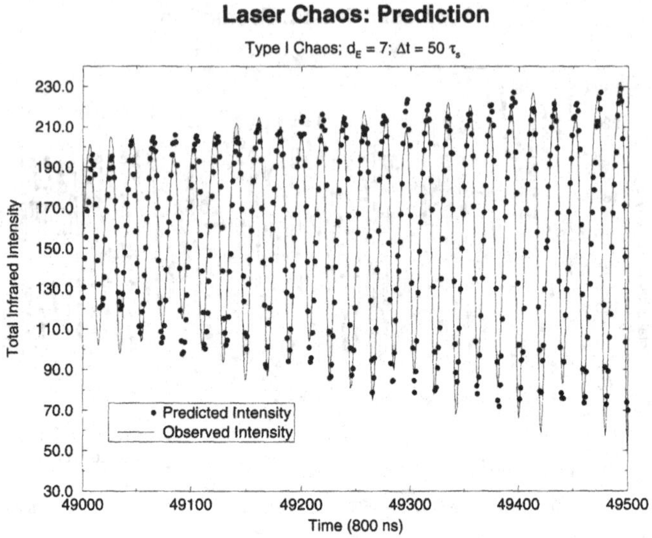

FIGURE 10.19.   Predicted and observed total infrared intensity for Type I chaos. The predictions are made using local linear maps in the reconstructed phase space of the attractor. The coefficients for the local maps are learned from the first 40,000 points of the data set, and then predictions are made starting with point 45,000. All predictions are made in $d_E = d_L = 7$. This figure shows the result of predicting ahead $\Delta t = 50\tau_s$. This is a region of phase space where the predictions are much better than would be expected from the values of the **global** Lyapunov exponents.

to evaluate any of these properties of the system strange attractor from **any** of the dynamical variables of the system, and that would constitute a quite stringent check on the results of the data analysis.

10.2   Chaotic Volume Fluctuations of the Great Salt Lake

The Great Salt Lake (GSL) is a shallow, salty body of water which drains an area of 90,000 km^2 in the Great Basin desert region of the United States. There are no outlet streams from the lake while input comes from rainfall and runoff from nearby mountains. The volume of the GSL has been measured by various methodologies every 15 days since 1847. When the analysis reported here was performed, 3463 data points (144 years) were available From a past calibration connecting height to lake volume, the volume of the lake is inferred from the average height at some fiducial locations. The average height thus serves as a proxy measurement for the GSL volume. From these data a time series reported in units of 10^7 acre-feet for the GSL

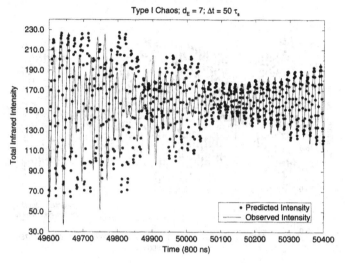

Laser Chaos: Prediction

FIGURE 10.20.   Predicted and observed total infrared intensity for Type I chaos. The predictions are made using local linear maps in the reconstructed phase space of the attractor. The coefficients for the local maps are learned from the first 40,000 points of the data set, and then predictions are made starting with point 45,000. All predictions are made in $d_E = d_L = 7$. This figure shows the result of predicting ahead $\Delta t = 50\tau_s$. This is a region of phase space were the predictions are much worse than when we tried to predict ahead only $\Delta t = 10\tau_s$.

volume has been compiled by Lall and Sangoyomi [San93]. The time series for GSL volume is shown in Figure 10.26.

The interest in the GSL volume is at least threefold:

i numerous economic assets are located along on near the lake. The brochure for the Great Salt Lake produced by the United States Geological Survey shows a resort area which has been severely damaged by variations in the lake's periphery associated with the volume change. In the early 1980s the volume of the lake showed a steady rise. Since the lake is so shallow, this was accompanied by an impressive increase in the lake extent. These rising waters threatened the Union Pacific Railroad, the Salt Lake City airport, the roadway of Interstate 80, and other industry such as the resort. To protect these assets the State of Utah spent nearly $150 million to develop a pumping station which was to remove lake water and send it to other areas for evaporation. This pumping plant was completed as the waters of the GSL began to recede. Knowledge of the GSL attractor such as we will discuss in a moment would have permitted a quite different planning process. Perhaps this would have resulted in expenditure of far fewer resources to combat a problem which clearly solved itself.

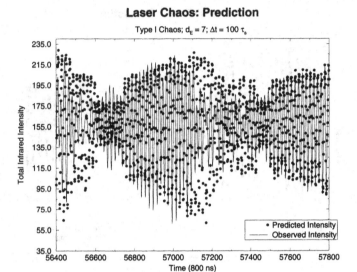

FIGURE 10.21.   Predicted and observed total infrared intensity for Type I chaos. The predictions are made using local linear maps in the reconstructed phase space of the attractor. The coefficients for the local maps are learned from the first 40,000 points of the data set, and then predictions are made starting with point 45,000. All predictions are made in $d_E = d_L = 7$. This figure shows the result of predicting ahead $\Delta t = 100\tau_s$. The predictions are much worse than above since we are trying to predict quite a bit beyond $\frac{\tau_s}{\lambda_1}$.

> ii the GSL volume integrates over a vast geographic region and numerous detailed small space and time scale physical processes such as local inflow, daily evaporation, and rainfall, etc. The GSL volume serves as a sensor for **climate change** which is rather insensitive to high frequency or small spatial scale variation. This is in striking contrast to point measurements which are typically very sensitive to short spatial scale phenomena even if they somehow accurately integrate over time variations.

> iii The GSL volume is also the longest geophysical time series known to me which is constituted of actual measurements made on the spot by human observers. Other long time series are inferred from alternative records or come from phenomena which predate human observation. They may be accurate, but then again, perhaps not. This long baseline time series may well serve to establish natural climate variations against which one may hope to see any effects on global climate change induced by human activities.

From the GSL volume time series we evaluate the Fourier spectrum of the data, and show it in Figure 10.27. One can see clearly visible annual

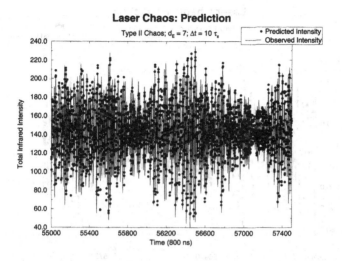

FIGURE 10.22.   Predicted and observed total infrared intensity for Type II chaos. The predictions are made using local linear maps in the reconstructed phase space of the attractor. The coefficients for the local maps are learned from the first 40,000 points of the data set, and then predictions are made starting with point 45,000. All predictions are made in $d_E = d_L = 7$. This figure shows the result of predicting ahead $\Delta t = 10\tau_s$.

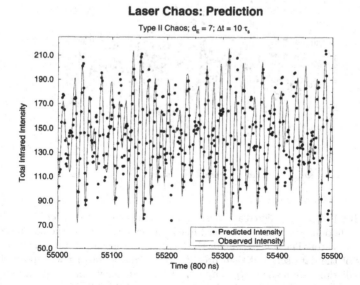

FIGURE 10.23.   An enlargement of the predictions for $\Delta t = 10\tau_s$ in Type II chaos.

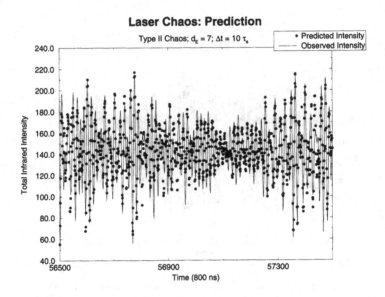

FIGURE 10.24.   An enlargement of the predictions for $\Delta t = 10\tau_s$ in Type II chaos.

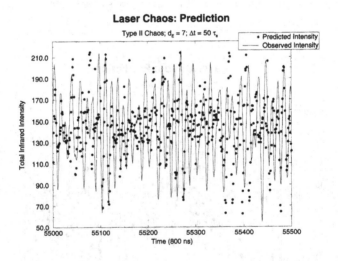

FIGURE 10.25.   Predicted and observed total infrared intensity for Type II chaos. The predictions are made using local linear maps in the reconstructed phase space of the attractor. The coefficients for the local maps are learned from the first 40,000 points of the data set, and then predictions are made starting with point 45,000. All predictions are made in $d_E = d_L = 7$. This figure shows the result of predicting ahead $\Delta t = 50\tau_s$. Clearly we have exceeded the predictability horizon in trying to predict this far ahead.

Great Salt Lake Volume

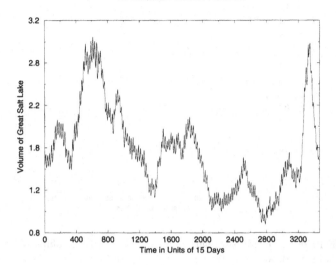

FIGURE 10.26.   The time series of the Great Salt Lake volume in units of 10^7 acre feet. The time index begins in 1847 and continues with measurements every 15 days.

Power Spectrum

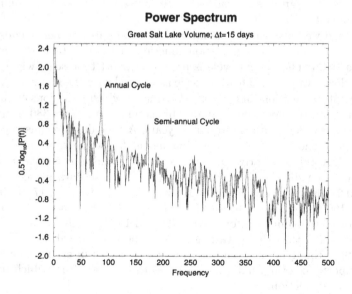

FIGURE 10.27.   The Fourier power spectrum of the GSL time series. The units on the frequency axis are $\frac{1}{85}$ years.

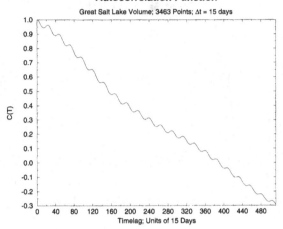

Autocorrelation Function

Great Salt Lake Volume; 3463 Points; Δt = 15 days

FIGURE 10.28.   The autocorrelation function for the GSL volume. The first zero crossing which indicates coordinates $v(n)$ and $v(n+T)$, which are linearly independent, is $T \approx 400$. The first local minimum of the autocorrelation function happens to be near the first minimum of the average mutual information.

and semi-annual cycles in the data. These periodic behaviors do not so dominate the signal as to lead us to see the GSL volume as an annual and semi-annual cycle with some small background chaos or noise imposed on it. Instead we have some dynamics driven, no doubt, by periodic external solar forcing. The autocorrelation function of these data is displayed in Figure 10.28. The annual cycle is present as short time scale wiggles on a much slower variation. The time scale associated with the first zero crossing of the autocorrelation function gives a time at which variations of the GSL volume are **linearly** decorrelated. This is quite long: $T \approx 380$ in units of 15 days, or a delay time of about 16 years. A time delay reconstruction of the phase space using this large time as the lag would quite likely reveal little, since this is clearly too long for any dynamical correlations in this system to persist. In sharp contrast $I(T)$ for these data as seen in Figure 10.29 shows a clear first minimum near $T \approx 11 - - - 17$ or times of about 160 to 250 days. The striking difference between the first zero of the autocorrelation function for these data and the first minimum of $I(T)$ is a clear signal that these statistics about the data reveal quite disparate aspects of the dynamics. We have argued at length in earlier chapters for the importance of mutual information as the statistic from which to make nonlinear deductions.

Using a time lag $T = 12$ we use the whole data set to discover the percentage of global false nearest neighbors shown in Figure 10.30. There is a sharp drop to zero at $d_E = 4$ after which the percentage of false

Average Mutual Information

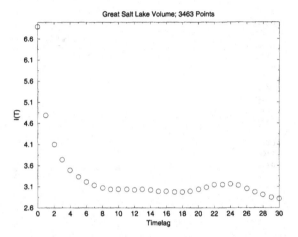

FIGURE 10.29.   The average mutual information for the GSL time series. It has a broad first minimum in the region of $T \approx 12 - -17$ in units of 15 days. As a prescription we choose a time lag in the region of the first minimum for use in reconstructing a multivariate state space using time lagged version of the GSL volume.

False Nearest Neighbors

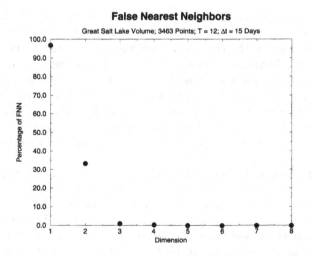

FIGURE 10.30.   Global false nearest neighbors for the GSL volume. At $d_E = 4$ the percentage of false neighbors goes to zero and stays there. This figure uses $T = 12$ for multivariate state space vectors; using $T = 17$ does not change the conclusion.

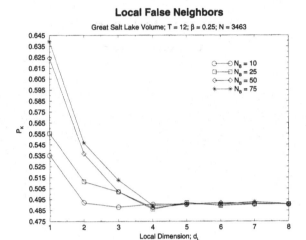

FIGURE 10.31.   Local false nearest neighbors for the GSL volume. The percentage of bad predictions P_K is given as a function of the local dimension d_L and the number of neighbors N_B used in making the neighborhood to neighborhood predictions. We show results for $N_B = 10, 25, 50,$ and 75. The predictions are deemed bad if they deviate by a fraction $\beta = 0.25$ of the size of the attractor in time T. As one varies β, the curve of bad predictions shifts up or down in P_K, but still becomes independent of N_B and d_L at $d_L = 4$.

neighbors remains zero. This provides evidence that we are dealing with a low dimensional system. The strength of this conclusion is enhanced when we see that varying the time delay over the range noted changes nothing but the detailed number of false neighbors at dimensions $d < d_E$. For $d \geq 4$ the percentage of false nearest neighbors is always zero. If one were to have used the time delay $T \approx 400$ suggested by the autocorrelation statistic a very high dimension would be predicted for the embedding. This is clearly a numerical artifact of the very large and inappropriate T and comes from the intrinsic instabilities in the system which are associated with the presence of chaos.

The data from the volume of the Great Salt Lake showed that globally false nearest neighbors become zero at $d_E = 4$. In Figure 10.31 we display the result of examining this data locally using local false nearest neighbors. The percentage of bad predictions seen in Figure 10.31 becomes independent of number of neighbors N_B and of the local dimension at $d_L = 4$, telling us that this attractor has four dynamical degrees of freedom. As indicated this means that models we would make of this process should have local four-dimensional dynamics regardless of the dimension of the overall space in which the model is built.

The final characteristic of the data we determine is the spectrum of Lyapunov exponents. Using the procedure described earlier we arrive at

Average Lyapunov Exponents

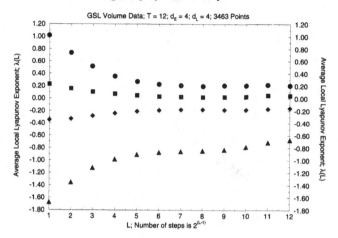

GSL Volume Data; T = 12; d_E = 4; d_L = 4; 3463 Points

FIGURE 10.32.   The average local Lyapunov exponents evaluated in $d_E = d_L = 4$ using a time delay of $T = 12$. The local exponents are evaluated at 750 locations around the attractor for $2^{(L-1)}$ steps forward from that location; we use $L = 1, 2, \ldots, 12$. The large L limit of these average exponents are the global Lyapunov exponents. Here we have $\lambda_1 > 0$ which is a critical indicator of chaos, and $\lambda_2 = 0.0$ which tells us a set of differential equations is at the source of our data.

the results shown in Figure 10.32. We see that there is again a single zero exponent indicating that we have a differential equation describing the dynamics. The Lyapunov dimension is determined by the values of the λ_a: $\lambda_1 = 0.17, \lambda_2 = 0.0, \lambda_3 = -0.14$, and $\lambda_4 = -0.65$ in inverse units of $\tau_s = 15$ days. This yields $D_L \approx 3.05$. Since the predictability time is about λ_1^{-1} this means that models for the processes producing fluctuations in the volume of the Great Salt Lake should allow prediction for about three months from any given time before the intrinsic instabilities of the system mask any such ability. As the dimension is nearly three, we anticipate that the structure of the attractor might be apparent if displayed in three dimensions $[v(t), v(t + T), v(t + 2T)]$. In Figure 10.33 we show this, and accepting the fact that some overlaps remain at $d_E = 3$, we see a rather well unfolded geometry. The annual cycle is associated with motion around the "spool" of the attractor, and the longer term evolution is along the length of the spool.

We do not have a set of four differential equations to model the GSL dynamics, but we can nonetheless use our knowledge of the dynamics in $d = 4$ to model and predict the evolution of the GSL volume. Here we confine ourselves to the use of local linear neighborhood to neighborhood maps.

In Figure 10.34 we display the result of predicting one step ahead for the 100 time points after time index 3301; April 24, 1985 to June 2, 1989. The

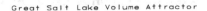

Great Salt Lake Volume Attractor

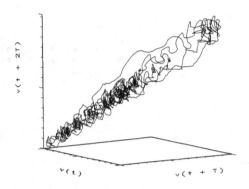

v (t + 2T)

v (t)   v (t + T)

FIGURE 10.33.   The strange attractor for the GSL volume as seen in time delay $[v(t), v(t+T), v(t+2T)]$ space with $T = 12$. The Lyapunov dimension of the GSL volume attractor is $D_L \approx 3.05$, so it is nearly unfolded in this picture.

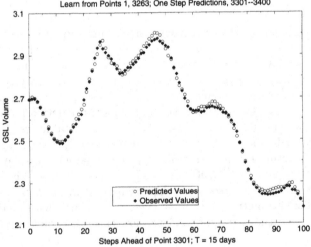

Predictions of GSL Volume

Learn from Points 1, 3263; One Step Predictions, 3301--3400

FIGURE 10.34.   Predictions using a local linear predictor for the GSL volume along with observed values. These predictions have been made one time step or 15 days ahead. The local linear maps for each region of the attractor were learned from the points 0 to 3263 of the data. Predictions were then made from points 3301 through 3400.

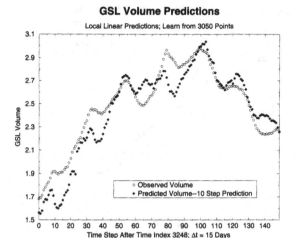

FIGURE 10.35.   Predictions using a local linear predictor for the GSL volume along with observed values. These predictions have been made 10 time steps or 150 days ahead. The local linear maps for each region of the attractor were learned from the points 0 to 3050 of the data. Predictions were then made from points 3248 through 3398.

data used to determine the local linear maps was taken from the beginning of the time series up to point 3263. It is clear that the ability to predict one step (15 days) ahead is very high. Next we determined the local linear maps from data starting at the beginning of the data set and using points up to index 3050. In Figure 10.35 we display the result of predicting 10 time steps ahead, that is 150 days or about a half year, again starting with time index 3248. In Figure 10.36 we learned local linear maps from the first 3200 points of the data set, and then predicted ahead ten steps from time index 3348. These predictions are actually rather good, and, as we can see, the prediction ability degrades with the number of steps ahead one wishes to predict, and this is as it should be in a chaotic system. The main message in these predictions is that we can accurately follow the features of the data without elaborate computations by working in dimension four as our systematic analysis would suggest. Of course, it must be possible to devise even better predictors using more elaborate methods.

10.2.1   A Few Remarks About the Great Salt Lake Volume

This analysis has not gone as far as the analysis of the laser intensity fluctuations. The main missing ingredient is a sense, from some first principles, of the dynamical equations for the lake volume. Of course, when sets of dynamical equations are suggested, we have bounded the number of equations which need have dynamical significance to four, determined that they must

FIGURE 10.36.   Predictions using a local linear predictor for the GSL volume along with observed values. These predictions have been made 10 time steps or 150 days ahead. The local linear maps for each region of the attractor were learned from the points 0 to 3200 of the data. Predictions were then made from points 3348 through 3398.

be differential equations because one of the global Lyapunov exponents is zero, and provided several invariants which must be reproduced by any model. These constraints should prove quite useful in making and testing realistic models for these phenomena.

A less precise achievement has been the identification and classification of a long, well measured geophysical data set which may prove useful in characterizing the natural variation of climate and thus provide an important baseline against which to identify human caused climate change. The policy and mitigation issues which are then raised go far beyond the scope or competence of this book.

10.3   Chaotic Motion in a Fluid Boundary Layer

Our final example [AKG*94] comes from measurements of wall pressure made along the outside of a torpedo body which was "launched" under its own buoyancy from depths of a few hundred meters in a fresh water lake. The body was instrumented with pressure sensors along eight axial lines with two lines of sensors at 0°, two at 90°, etc. These lines of sensors were flush with the exterior of the body and stretched from about 10 cm to 50 cm back from the prow of the body. Measurements were made at one of the sensors before the body was launched thus providing data for the

ambient state associated with the lake and the measurement apparatus.
Two accelerometers were also utilized. One was forward and the other aft,
and both were within the torpedo body.

After release the steady motion of the body lasted about 10 s in each of
the many experiments made on this object, and the pressure at all locations
was sampled at 2^{16} Hz or $\tau_s = 15.26\,\mu s$. Data was collected for 5 to 6 seconds
at each sensor then filtered with a low pass filter having a sharp rolloff at
6400 Hz.

In steady motion the speed u of the vehicle was about 20 m/s, and the
thickness δ of the boundary layer around the body ranged from $\delta \approx 0.16$
cm to $\delta \approx 0.6$ cm depending on the distance from the prow. The natural
time scale of the fluid dynamics in the boundary layer is $\frac{\delta}{u} \approx 250\,\mu s$. Based
on u and δ, the Reynolds number for these flows was about 100,000. The
friction velocity u_τ, which is related to the stress on the wall, was $u_\tau \approx$
0.73 m/s, and this leads to a length scale defining what are called "wall
units" [Can81, Rob91] based on u_τ and the kinematic viscosity ν:

$$\frac{\nu}{u_\tau} \approx 2.6 \times 10^{-4}\text{cm}. \qquad (10.16)$$

In these wall units the size of each pressure sensor was about 1400.

A significant feature of boundary layer turbulent flows and other shear
flows is the appearance of coherent structures [Can81, Rob91] in both the
spanwise and streamwise directions. The spanwise structures are dominated
by horseshoe like vortex elements which are bent by the mean flow from being mostly spanwise to being stretched in the streamwise direction as well.
These horseshoe structures are large, about 150 to 200 wall units in size,
and their chaotic motions substantially influence the pressure fluctuations
on the sensors in a mid-range of frequencies f where

$$\frac{2\pi f\delta}{u_\tau} \approx 50, \qquad (10.17)$$

as reported in [FC91]. It has been estimated by Schewe [Sch83] that a
sensor size of about 20 or smaller in wall units is required to resolve the fine
structure in the boundary layer flow. This would imply that the very large
sensors used in this series of experiments average over the spatial degrees of
freedom with high wave number and thus are sensitive to fluctuations in the
influence of the coherent structures on the flow at the wall. A rough estimate
of the number of coherent structures influencing each of the pressure sensors
is about 10 to 12. We might well hope to see low dimensional chaos in these
pressure signals, if the data is clean.

We will examine here data from two sensors: the first is located reason-
ably far aft and is in a region where the turbulence in the boundary layer
would be well developed. This is called sensor B7 in the experiment. We
will look at data from this sensor when the body is resting at the bottom
of the lake before rising under its own buoyancy and also we look at data

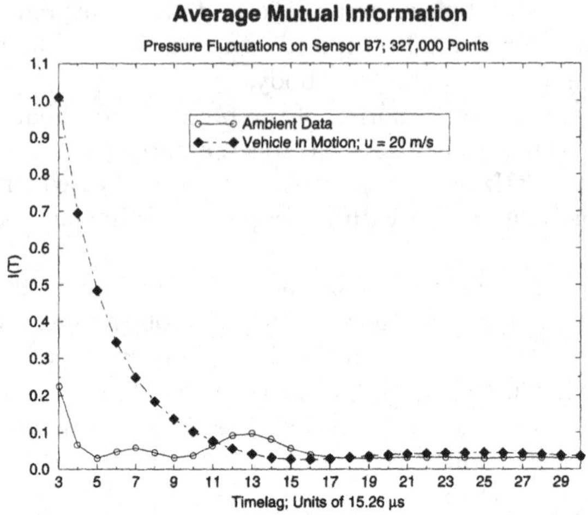

Average Mutual Information

Pressure Fluctuations on Sensor B7; 327,000 Points

FIGURE 10.37.   The average mutual information for wall pressure measurements from the sensor at station B7 both when the buoyant vehicle was at rest (open circles joined with a solid line) and when it was moving at $u \approx 20.8$ m/s (filled diamonds joined with a solid line). The first minimum for the ambient data is at $T = 5$ or about $76.30\,\mu$sec. The first minimum for the moving vehicle is near $T = 18$. The latter is characteristic of the fluid dynamics of the boundary layer.

from the same sensor while the body is in steady motion at about 22.8 m/s. The second sensor, called C5 in the experiment, is located in a region where the **transition** from laminar flow over the prow to turbulent flow rearward is taking place.

For sensor B7 we skip the time series and Fourier spectra taken while the vehicle was in motion or at rest. They reveal the usual complex time traces and broadband spectra. In Figure 10.37 we show the average mutual information for both the moving and rest states. Clearly the dynamics of pressure fluctuations along the wall of the body changed as the body moved through the water. The **ambient** pressure signal shows a minimum in $I(T)$ near $T = 5$ or about 75 μs while the signal from the same sensor when the body was moving shows a first minimum in $I(T)$ near $T = 18$ or 270 μs. The second is a signal of fluid dynamic origin, while the first, presumably the background, is of another origin. The analysis here, and the analysis of the signals from tens of other sensors on this body verify this conclusion. The data from the sensor C5 in the transition flow region yields an $I(T)$ as seen in Figure 10.38. Here the first minimum at $T = 16$ also shows these data to be associated with fluid motions. $T = 16$ corresponds to 240 μs.

Next we turn to the analysis of global false nearest neighbors for data from stations B7 and C5. For the ambient data we have a time delay of $T = 5$, and the false nearest neighbors result is seen in Figure 10.39. It

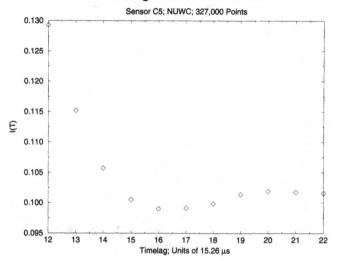

FIGURE 10.38.   The average mutual information from wall pressure measured at the station C5 when the buoyant vehicle was moving at $u \approx 22.8$ m/s. The first minimum is near $T = 16$.

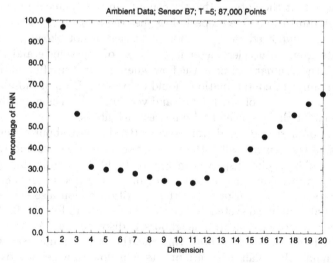

FIGURE 10.39.   Global false nearest neighbors for wall pressure data measured at station B7 when the buoyant vehicle was at rest. This is characteristic of high dimensional dynamics or "noise".

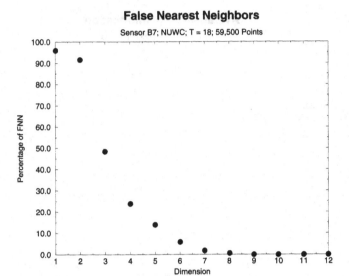

False Nearest Neighbors

Sensor B7; NUWC; T = 18; 59,500 Points

FIGURE 10.40.   The average mutual information for wall pressure data measured at station B7 (in the turbulent region of boundary layer flow) when the vehicle was moving at $u \approx 22.8$ m/s. There is strong evidence of low dimensional dynamics operating here.

seems clear that the data in this set come from a dynamical process with quite high dimension. One suspects it is just the noise in the sensors. That suspicion is supported by two kinds of data sets not shown here: measurements from an accelerometer in the nose of the vehicle and pressure sensors in the laminar regime of the flow where no pressure fluctuations except from ambient contamination should be present. Both of those kinds of data give time delays of order $T \approx 5$ and very high dimensional embedding dimensions for the absence of false nearest neighbors.

For the data set acquired from sensor B7 (located about 50 cm from the prow of the vehicle) while the vehicle was in motion, we find the false nearest neighbors plot displayed in Figure 10.40. It is clear that we now have a low dimensional attractor which can be captured in $d_E \approx 8$.

The data set from sensor C5 in the transition region also shows a low dimensional dynamical system in action. This is seen in Figure 10.41 where the percentage of false nearest neighbors is displayed and in an enlarged format in Figure 10.42. For both this and the previous data set we have already made the qualitative arguments why low dimensional dynamics could be present. Despite those arguments it is remarkable that within a fully turbulent (B7) flow or a transitional flow (C5) where the total number of active degrees of freedom is certainly large, we have a sensor which has carefully filtered out the small scale dynamics and exposed the primary

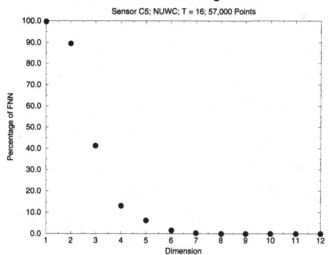

FIGURE 10.41. The average mutual information for wall pressure data measured at station C5 (in the transition region of boundary layer flow) when the vehicle was moving at $u \approx 22.8$ m/s. There is strong evidence of low dimensional dynamics operating here.

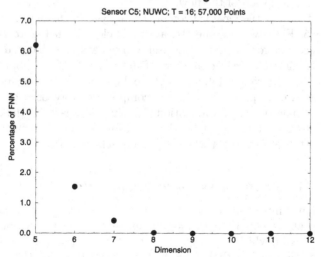

FIGURE 10.42. An expanded version of Figure 10.41 showing that global false nearest neighbors goes to zero at $d_E = 8$.

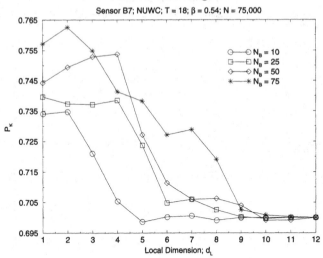

FIGURE 10.43.   Local false nearest neighbors for wall pressure data from station B7 when the vehicle was moving at $u \approx 22.8$ m/s. The evidence is not as clear here because of the high dimensions involved, but a local dynamics near dimension $d_L = 9$ appears to be selected by the data.

variations of the large scale coherent structures which play such a dominant role in the description of the flow.

We next look at the local version of the analysis of the data from sensors B7 and C5. First we examine B7, and it is clear from Figure 10.43 that $d_L = 9$ is selected for this data. It is useful to remark here that in dimensions as high as eight to ten the amount of data one needs to populate local regions of the attractor fully enough to be sure of the results of these local tests can be quite large. In our computations we used of order 10^5 data points in each case. From station C5, which is located in the transition region of the boundary layer flow, the result for local false nearest neighbors is displayed in Figure 10.44 where $d_L = 9$ is selected.

10.3.1   A Few Remarks About Chaotic Boundary Layers

This section is intermediate between the first two in the sense that we have some idea about equations of motion, namely, the dynamics of coherent vortex filaments, which could apply to the observed phenomena. The equations of motion are in that sense better known than for the Great Salt Lake dynamics, but far less well known than for the solid state laser. The observations here depend on the sensor having averaged over the smallest scale fluctuations in the fluid, and in that regard the situation is similar to

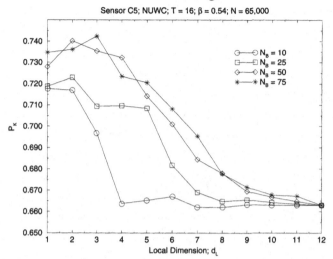

FIGURE 10.44.   Local false nearest neighbors for wall pressure data from station C5 when the vehicle was moving at $u \approx 22.8$ m/s. The evidence is not as clear here because of the high dimensions involved, but a local dynamics near dimension $d_L = 9$ appears to be selected by the data.

the spatial filtering of climate details by the Great Salt Lake volume measurements. There is clearly a long way to go from the analysis of these data to a quantitative set of vortex filament dynamical equations which reproduce the observations. One of the preliminary tasks would be to repeat the experiments in a more controlled setting and examine in a detailed fashion the dependence on sensor size of the conclusions about the dimension of the dynamics in each flow regime.

11
Estimating in Chaos: Cramér-Rao Bounds

Estimating parameters of the dynamics from observations of a noise contaminated time series or estimating the clean time series itself is a classical challenge in the analysis of signals [Tre68]. This chapter contributes to the analysis of one's ability to estimate a clean signal from observations of a noisy signal with the study of the case when the clean signal arises from a nonlinear dynamical system which exhibits chaotic evolution. We will argue that aspects of the very property of chaotic systems which makes them only slightly predictable and causes them to be nonperiodic, namely, the presence of positive Lyapunov exponents, also is connected with a Cramér-Rao bound which suggests an ability to estimate the clean time series with exponential accuracy. The bounds presented here give a rationale for the striking results on how well one can estimate clean data from noisy observations associated with manifold decomposition [Ham90, FS89, ABST93] as discussed in Chapter 7.

The results presented in this chapter were worked out with M. D. Richard and are based on many results contained in his PhD dissertation [Ric94]. The results were reported in the proceedings of a conference in 1993 [Aba93], but they have not been published in any detail. Because estimating in the presence of noise is an important signal processing matter, and because the possibilities are so much more striking when the source of the clean signal is nonlinear, we present some details of the arguments for the nonlinear dynamical version of the Cramér-Rao bounds.

11.1   The State Estimation Problem

We start with discrete time dynamics reached by finite time sampling $\mathbf{x}(n) = \mathbf{x}(t_0 + n\tau_s)$ or by stroboscopically looking at a Poincaré section of the orbit. In either case, the dynamics of the discrete time system satisfy an evolution equation

$$\mathbf{x}(n+1) = \mathbf{F}(\mathbf{x}(n)). \tag{11.1}$$

We assume that $\mathbf{F}$ is a diffeomorphism, which is a reasonable assumption in light of its origin. Indeed, when this assumption "fails", as would seem to be the case in stick-slip friction problems, it means that physically we fail to resolve the system on a short enough time scale to see the smooth behavior inherent in all real observations. Such nondifferentiable dynamics will require additional methods beyond what we are able to address.

In many real-world applications, one does not actually observe the d-dimensional state vector $\mathbf{x}(n)$ of the system, but only a p-dimensional vector $\mathbf{y}(n)$ related to $\mathbf{x}(n)$ by

$$\mathbf{y}(n) = \mathbf{h}(\mathbf{x}(n)) + \mathbf{v}(n), \tag{11.2}$$

where $\mathbf{h}(\bullet)$ is a memoryless but not necessarily invertible mapping from $R^d \to R^p$. $\mathbf{v}(n)$ is p-dimensional noise with probability density $p_v(\mathbf{v}(n))$. $\mathbf{v}(n)$ could come itself from another chaotic system, and $p_v(\mathbf{v})$ would be the invariant density for that system [ABST93]. For our considerations it is only necessary that the contamination $\mathbf{v}(n)$ have some well defined appropriately differentiable distribution function in p-dimensional state space. In the discussion that follows, we often assume that the time sequence $\{\mathbf{v}(i)\}$ is a discrete-time, Gaussian white-noise process. It is always taken to be independent of the state vectors $\mathbf{x}(n)$. We consider only the case in which $\mathbf{h}(\bullet)$ is the identity operator. Our results easily generalize for an arbitrary differentiable, memoryless transformation $\mathbf{h}(\bullet)$, and the Gaussianity of the contamination is not essential.

The particular problem we discuss is that of estimating the value of the state $\mathbf{x}(n)$ at a single time n given a set of $N + M + 1$ noisy, sequential observations $\boldsymbol{Y}_M^N$ where

$$\boldsymbol{Y}_M^N = \{\mathbf{y}(-M), \mathbf{y}(-M+1), \ldots, \mathbf{y}(N-1), \mathbf{y}(N)\},$$
$$\mathbf{y}(k) = \mathbf{x}(k) + \mathbf{v}(k). \tag{11.3}$$

Although we place no restrictions on n, M, and N, our principal interest is when $-M \ll n \ll N$, that is, we are interested in estimating the state $\mathbf{x}(n)$ when we have available observations of $\mathbf{y}(n)$ from the distant past as well as from the far future of the "time" n. In the sequel, we let $\hat{\boldsymbol{x}}(n, \boldsymbol{Y}_M^N)$ denote an estimator of $\mathbf{x}(n)$ and our focus will be on obtaining a lower bound on $\boldsymbol{P}(\hat{\boldsymbol{x}}(n, \boldsymbol{Y}_M^N))$, the error covariance matrix of $\mathbf{x}(n)$, defined as

$$\boldsymbol{P}(\hat{\boldsymbol{x}}(n, \boldsymbol{Y}_M^N)) \equiv E\left[(\hat{\boldsymbol{x}}(n, \boldsymbol{Y}_M^N) - \mathbf{x}(n))(\hat{\boldsymbol{x}}(n, \boldsymbol{Y}_M^N) - \mathbf{x}(n))^T\right] \tag{11.4}$$
$$= \int (\hat{\boldsymbol{x}}(n, \boldsymbol{Y}_M^N) - \mathbf{x}(n))(\hat{\boldsymbol{x}}(n, \boldsymbol{Y}_M^N) - \mathbf{x}(n))^T p(\boldsymbol{Y}_M^N | \mathbf{x}(n))\, d\boldsymbol{Y}_M^N$$

where $p(\boldsymbol{Y}_M^N|\mathbf{x}(n))$ is the "likelihood function" or conditional probability density of the observations $\boldsymbol{Y}_M^N$ given the state $\mathbf{x}(n)$.

The likelihood function has a fundamental role in estimation theory. The value of $\mathbf{x}(n)$ that maximizes this for a given set of observations $\boldsymbol{Y}_M^N$ is called the "Maximum Likelihood" estimate of the state $\mathbf{x}(n)$ at time n. In the next section, we show how one can use the likelihood function to derive a lower bound on the error covariance matrix $\boldsymbol{P}(\hat{\boldsymbol{x}}(n,\boldsymbol{Y}_M^N))$. This lower bound, known as the Cramér-Rao bound, is the most widely used error bound for parameter estimation problems. An interesting aspect of this bound is that if an estimator has performance which actually achieves the bound, the estimator is also the Maximum Likelihood estimator [Tre68]. The bound only applies to **unbiased** estimators which satisfy

$$E(\hat{\boldsymbol{x}}(n,\boldsymbol{Y}_M^N)|\mathbf{x}(n)) = \int \hat{\boldsymbol{x}}(n,\boldsymbol{Y}_M^N)p(\boldsymbol{Y}_M^N|\mathbf{x}(n))\,d\boldsymbol{Y}_M^N = \mathbf{x}(n). \qquad (11.5)$$

11.2   The Cramér-Rao Bound

The general form of the Cramér-Rao bound on $\boldsymbol{P}(\hat{\boldsymbol{x}}(n,\boldsymbol{Y}_M^N))$ is [Tre68]

$$\boldsymbol{P}(\hat{\boldsymbol{x}}(n,\boldsymbol{Y}_M^N)) \geq \boldsymbol{J}^{-1}(\mathbf{x}(n)), \qquad (11.6)$$

where $\boldsymbol{J}(\mathbf{x}(n))$, known as the **Fisher information matrix**, is

$$\boldsymbol{J}(\mathbf{x}(n)) = E\left\{\left(\left[\frac{\partial \log p(\boldsymbol{Y}_M^N|\mathbf{x}(n))}{\partial \mathbf{x}(n)}\right]\left[\frac{\partial \log p(\boldsymbol{Y}_M^N|\mathbf{x}(n))}{\partial \mathbf{x}(n)}\right]^T\right)\right\} \qquad (11.7)$$

$$= \int d\boldsymbol{Y}_M^N \frac{\partial \log p(\boldsymbol{Y}_M^N|\mathbf{x}(n))}{\partial \mathbf{x}(n)}\left[\frac{\partial \log p(\boldsymbol{Y}_M^N|\mathbf{x}(n))}{\partial \mathbf{x}(n)}\right]^T p(\boldsymbol{Y}_M^N|\mathbf{x}(n)) \quad .$$

The (α,β) component of the Fisher information matrix is

$$\boldsymbol{J}_{\alpha,\beta}(\mathbf{x}(n)) = \int \frac{\partial \log p(\boldsymbol{Y}_M^N|\mathbf{x}(n))}{\partial x_\alpha(n)}\frac{\partial \log p(\boldsymbol{Y}_M^N|\mathbf{x}(n))}{\partial x_\beta(n)}p(\boldsymbol{Y}_M^N|\mathbf{x}(n))\,d\boldsymbol{Y}_M^N.$$
$$(11.8)$$

where $\alpha,\beta = 1,2,\ldots,d$ and $x_\alpha(n)$ denotes the αth component of $\mathbf{x}(n)$.

The derivation of the Cramér-Rao bound starts from the requirement that the observations be unbiased which we write as

$$\int \left[\hat{\boldsymbol{x}}(n,\boldsymbol{Y}_M^N) - \mathbf{x}(n)\right]p(\boldsymbol{Y}_M^N|\mathbf{x}(n))\,d\boldsymbol{Y}_M^N = 0. \qquad (11.9)$$

Differentiating this expression with respect to $x_\alpha(n)$, we have

$$0 = -\delta_{\alpha\beta} + \int d\boldsymbol{Y}_M^N \frac{\partial \log p(\boldsymbol{Y}_M^N|\mathbf{x}(n))}{\partial x_\alpha(n)}p(\boldsymbol{Y}_M^N|\mathbf{x}(n))[\hat{x}(n,\boldsymbol{Y}_M^N)_\beta - x_\beta(n)].$$
$$(11.10)$$

or

$$\delta_{\alpha\beta} = \int d\mathbf{Y}_M^N \{ \frac{\partial \log p(\mathbf{Y}_M^N | \mathbf{x}(n))}{\partial x_\alpha(n)} \sqrt{p(\mathbf{Y}_M^N | \mathbf{x}(n))} \} \{ \sqrt{p(\mathbf{Y}_M^N | \mathbf{x}(n))}$$
$$\times [\hat{x}(n, \mathbf{Y}_M^N)_\beta - x_\beta(n)] \}, \tag{11.11}$$

and using the Schwartz inequality on this expression we arrive at

$$\delta_{\alpha\beta} \leq \left[\int d\mathbf{Y}_M^N p(\mathbf{Y}_M^N | \mathbf{x}(n)) \frac{\partial \log p(\mathbf{Y}_M^N | \mathbf{x}(n))}{\partial x_\alpha} \frac{\partial \log p(\mathbf{Y}_M^N | \mathbf{x}(n))}{\partial x_\gamma} \right]$$
$$\times \left[\int d\mathbf{Y}_M^N p(\mathbf{Y}_M^N | \mathbf{x}(n)) [\hat{x}(n, \mathbf{Y}_M^N)_\gamma - x_\gamma(n)] \right.$$
$$\times [\hat{x}(n, \mathbf{Y}_M^N)_\beta - x_\beta(n)] \Big]$$
$$= \left[\mathbf{P}(\hat{x}(n, \mathbf{Y}_M^N)) \cdot \mathbf{J}(\mathbf{x}(n)) \right]_{\alpha\beta}. \tag{11.12}$$

This we recognize as the Cramér-Rao bound as stated earlier. The expression for the Fisher information matrix expression simplifies when we use the whiteness of the additive observation noise $\mathbf{v}(n)$ and the explicit dependence of the likelihood function $p(\mathbf{Y}_M^N | \mathbf{x}(n))$ on $\mathbf{x}(n)$. In this instance

$$p(\mathbf{Y}_M^N | \mathbf{x}(n)) = \prod_{i=-M}^{N} p_v(\mathbf{y}(i) - \mathbf{x}(i)), \tag{11.13}$$

where as noted earlier p_v is the probability density function for each noise term $\mathbf{v}(n)$. We can express the likelihood function as a product of terms since the system is deterministic and invertible, and the state $\mathbf{x}(i)$ at any time i is uniquely determined by the state $\mathbf{x}(n)$ at time n by $\mathbf{x}(i) = \mathbf{F}^{i-n}(\mathbf{x}(n))$. Since the noise is white, the observation $\mathbf{y}(i)$ conditioned on $\mathbf{x}(i)$ is independent of all other observations, and because the noise is additive, the conditional density $p(\mathbf{y}(i) | \mathbf{x}(i))$ has the form

$$p(\mathbf{y}(i) | \mathbf{x}(i)) = p_v(\mathbf{y}(i) - \mathbf{x}(i)), \tag{11.14}$$

and from this Equation (11.13) follows.

Since $\mathbf{x}(i) = \mathbf{F}^{i-n}(\mathbf{x}(n))$, we can express Equation (11.14) as

$$p(\mathbf{y}(i) | \mathbf{x}(i)) = p(\mathbf{y}(i) | \mathbf{F}^{i-n}(\mathbf{x}(n))) = p_v(\mathbf{y}(i) - \mathbf{F}^{i-n}(\mathbf{x}(n))) \tag{11.15}$$

and thus can write our expression for $p(\mathbf{Y}_M^N | \mathbf{x}(n))$ as

$$p(\mathbf{Y}_M^N | \mathbf{x}(n)) = \prod_{i=-M}^{N} p_v(\mathbf{y}(i) - \mathbf{F}^{i-n}(\mathbf{x}(n))). \tag{11.16}$$

The logarithm of this enters $\mathbf{J}(\mathbf{x}(n))$ and is

$$\log p(\mathbf{Y}_M^N|\mathbf{x}(n)) = \sum_{i=-M}^{N} \log p_v(\mathbf{y}(i) - \mathbf{F}^{i-n}(\mathbf{x}(n))), \qquad (11.17)$$

which gives

$$\frac{\partial \log p(\mathbf{Y}_M^N|\mathbf{x}(n))}{\partial x_\alpha(n)} = \sum_{l=-M}^{N} \sum_{\gamma=1}^{d} \sigma_\gamma(\mathbf{y}(l) - \mathbf{x}(l)) Q_{\gamma\alpha}^l(\mathbf{x}(n)), \qquad (11.18)$$

where

$$\sigma_\gamma(\mathbf{w}) = -\frac{\partial \log p_v(\mathbf{w})}{\partial w_\gamma},$$

$$Q_{\gamma\alpha}^l(\mathbf{x}(n)) = \frac{\partial F_\gamma^{l-n}(\mathbf{x}(n))}{\partial x_\alpha}, \qquad (11.19)$$

and $F_\gamma^{l-n}(\mathbf{x}(n))$ denotes the γth component of $\mathbf{F}^{l-m}(\mathbf{x}(n))$. Using the chain rule we can also express $Q_{\gamma\alpha}^l(\mathbf{x}(n))$ as

$$Q_{\gamma\alpha}^l(\mathbf{x}(n)) = [\mathbf{DF}(\mathbf{x}(l-1)), \cdots, \mathbf{DF}(\mathbf{x}(n+1)) \cdot \mathbf{DF}(\mathbf{x}(n))]_{\gamma\alpha}, \quad (11.20)$$

for $l \geq n+2$. For $l \leq n-2$, this becomes

$$Q_{\gamma\alpha}^l(\mathbf{x}(n)) = [\mathbf{DF}^{-1}(\mathbf{x}(l+1)), \cdots, \mathbf{DF}(\mathbf{x}(n-1)) \cdot \mathbf{DF}^{-1}(\mathbf{x}(n))]_{\gamma\alpha}, \quad (11.21)$$

Similar expressions hold for $n-1 \leq l \leq n+1$. In these expressions, $\mathbf{DF}(\mathbf{x})$ is the $d \times d$ Jacobian matrix

$$\mathbf{DF}(\mathbf{x}) = \frac{\partial \mathbf{F}(\mathbf{x})}{\partial \mathbf{x}}, \qquad (11.22)$$

$\mathbf{DF}^{-1}(\mathbf{x})$ is the $d \times d$ Jacobian matrix of $\mathbf{F}^{-1}(\mathbf{x})$.

The Fisher information matrix now becomes

$$\mathbf{J}(\mathbf{x}(n)) = \sum_{k,l=-M}^{N} [\mathbf{Q}^l(\mathbf{x}(n))]^T \mathbf{S}(l,k) \mathbf{Q}^k(\mathbf{x}(n)), \qquad (11.23)$$

and the matrix $\mathbf{S}(l,k)$ has components

$$S(l,k)_{\alpha\beta} = \int \prod_{k,l=-M}^{N} \sigma_\alpha(\mathbf{y}(l) - \mathbf{x}(l)) \sigma_\beta(\mathbf{y}(k) - \mathbf{x}(k)) p(\mathbf{Y}_M^N|\mathbf{x}(n)) d\mathbf{Y}_M^N.$$

$$(11.24)$$

This decomposition effectively decouples the dependence of $\mathbf{J}(\mathbf{x}(n))$ on the statistics of the noise $\mathbf{v}(n)$ and dynamics of the system $\mathbf{F}$. The noise statistics are captured in the matrix $\mathbf{S}(l,k)$, while the system dynamics resides in the matrices $\mathbf{Q}^l(\mathbf{x}(n))$.

Some perspective on these formulae comes from examining the case where $p_v(\mathbf{v}(n))$ is not only independently and identically distributed but Gaussian as well. In this case,

$$\sigma_\alpha(\mathbf{y}(l) - \mathbf{x}(l)) = \sum_{\beta=1}^{d} (\mathbf{R}^{-1})_{\alpha\beta}(\mathbf{y}(l) - \mathbf{x}(l))_\beta, \qquad (11.25)$$

where $\mathbf{R}$ is the covariance matrix of the noise

$$E[\mathbf{v}(k)\mathbf{v}(l)^T] = \delta_{kl}\mathbf{R}. \qquad (11.26)$$

The matrix $\mathbf{S}(l, k)$ is then given by

$$\mathbf{S}(l, k) = \mathbf{R}^{-1}\delta_{lk}. \qquad (11.27)$$

If, for further simplification, the noise is taken to be isotropic in state space, namely,

$$\mathbf{R} = \nu^2 \mathcal{I}, \qquad (11.28)$$

where $\mathcal{I}$ is the $d \times d$ identity matrix, then

$$\mathbf{S}(l, k) = \mathcal{I} \, \frac{1}{\nu^2}\delta_{kl}. \qquad (11.29)$$

The Fisher information matrix for this case simplifies to

$$\mathbf{J}(\mathbf{x}(n)) = \frac{1}{\nu^2} \sum_{k=-M}^{N} \boldsymbol{Q}^k(\mathbf{x}(n))^T \cdot \boldsymbol{Q}^k(\mathbf{x}(n)). \qquad (11.30)$$

At this stage we can make a connection between the local and global Lyapunov exponents of the dynamical system $\mathbf{F}$ and the Fisher information matrix. The local Lyapunov exponents are the logarithm of the eigenvalues of the Oseledec matrix [Ose68, ER85, ABK91]

$$\left\{\mathbf{Q}^L(\mathbf{x})^T \cdot \mathbf{Q}^L(\mathbf{x})\right\}^{\frac{1}{2L}}, \qquad (11.31)$$

where we recognize $\mathbf{Q}^L(\mathbf{x})$ as the $\mathbf{DF}^L(\mathbf{x})$ we used earlier.

We see that the Fisher information matrix consists of a sum of partial products of the infinite product of Jacobian matrices which determines the global Lyapunov exponents of the dynamical system whose orbit we are trying to estimate from noisy observations. It is clear from this that knowledge of the local and global Lyapunov exponents for the dynamics determine properties of $\mathbf{J}(\mathbf{x}(n))$.

11.3 Symmetric Linear Dynamics

We begin the study of the connection between the Fisher information matrix and Lyapunov exponents with a linear dynamical system [Ric94]

$$\mathbf{x}(n+1) = \mathbf{G} \cdot \mathbf{x}(n), \tag{11.32}$$

where the $d \times d$ matrix $\mathbf{G}$ is symmetric and has no null eigenvalues. This may seem like a physically nonsensical system, since if any of the eigenvalues of $\mathbf{G}$ are greater than unity the system is unstable. In the engineering community, such systems are often used to model unstable processes or plants that must be stabilized. Although linear systems play an important role in the traditional engineering community, rarely are deterministic, linear systems considered because of their trivial dynamics. Instead, a small driving noise component is generally included in the righthandrighthand side of the above equation. In any case we shall see that the consideration of this case illuminates our subsequent discussion where in a **nonlinear** system positive and negative global Lyapunov exponents coexist in any chaotic motions of the system. The sum of the exponents is negative, representing physical dissipation, and the nonlinearity folds the instability directions embodied in the positive Lyapunov exponents back onto the compact attractor.

In the eigenbasis of $\mathbf{G}$ we have

$$\mathbf{G} = \begin{bmatrix} e^{\lambda_1} & 0 & \cdots & 0 \\ 0 & e^{\lambda_2} & & 0 \\ \vdots & & \ddots & \vdots \\ 0 & 0 & \cdots & e^{\lambda_d} \end{bmatrix}, \tag{11.33}$$

where the eigenvalues $\{e^{\lambda_i}\}$ are written in this form to conform with the definition of Lyapunov exponents λ_a.

For this simple model the Jacobian matrix $\mathbf{DF}(\mathbf{x}(n)) = \mathbf{G}$ and is independent of the state space location $\mathbf{x}$. The Fisher information matrix is given by

$$\begin{aligned} \mathbf{J}(\mathbf{x}(n)) &= \frac{1}{\nu^2} \sum_{l=-M}^{N} \mathbf{G}^{2(l-n)} \\ &= \frac{1}{\nu^2} \begin{bmatrix} S_1 & 0 & \cdots & 0 \\ 0 & S_2 & & 0 \\ \vdots & & \ddots & \vdots \\ 0 & 0 & \cdots & S_d \end{bmatrix}, \end{aligned} \tag{11.34}$$

where

$$S_a = \sum_{l=-M}^{N} e^{2\lambda_a(l-n)}$$

$$= e^{-2n\lambda_a}\frac{e^{-2M\lambda_a} - e^{(2N+2)\lambda_a}}{1 - e^{2\lambda_a}}. \tag{11.35}$$

The Cramér-Rao bound for this system is given by

$$\mathbf{P}(\hat{\mathbf{x}}(n)) \geq \mathbf{J}^{-1}(\mathbf{x}(n)) = \nu^2 \begin{bmatrix} S_1^{-1} & 0 & \cdots & 0 \\ 0 & S_2^{-1} & & 0 \\ \vdots & & \ddots & \vdots \\ 0 & 0 & \cdots & S_d^{-1} \end{bmatrix}. \tag{11.36}$$

We are interested in the behavior of the bound as the number of observations becomes large while n is held fixed $= M \ll n \ll N$. This corresponds to the bound on one's ability to estimate $\mathbf{x}(n)$ when the number of observations both before and after time n increases. The behavior of S_a^{-1} in this situation is given by the following possibilities:

i If $\lambda_a > 0$, then the term $e^{-2M\lambda_a}$ is small and

$$S_a^{-1} \approx e^{(2n-2)\lambda_a}(e^{2\lambda_a} - 1)e^{-2N\lambda_a}. \tag{11.37}$$

In other words the bound on error variance for the corresponding component of the state vector goes to zero exponentially rapidly with increasing observations from the past and at a rate governed by the Lyapunov exponent λ_a.

ii If $\lambda_a < 0$, then the situation is reversed, and $e^{2N\lambda_a}$ is very small,

$$S_a^{-1} \approx e^{2n\lambda_a}(1 - e^{2\lambda_a})e^{-2M|\lambda_a|}. \tag{11.38}$$

Again the bound goes to zero exponentially rapidly; this time the observations from the future are essential.

iii If $\lambda_a \equiv 0$, then

$$S_a^{-1} = \frac{1}{M + N + 1}. \tag{11.39}$$

This means the error variance for the corresponding component of $\mathbf{x}(n)$ is bounded below by the reciprocal of the number of observations times the variance ν^2 of the contaminations.

The last case is the conventional one found in the literature [Tre68]. It leads to the familiar result that with $M + N + 1$ observations the error due to noise can be reduced to

$$\text{error} \approx \frac{\text{noise level}}{\sqrt{M + N + 1}}. \tag{11.40}$$

This is consistent with the remarks made in the chapter on signal separation. The distinctions among the three cases just cited is the critical result.

When the system is stable, the Lyapunov exponents are all negative, and the bound on the error variance drops to zero along each component of the state vector as the number of past observations goes to infinity. In such a case the only allowed states of the system are fixed points, and the orbits are really quite uninteresting.

Similarly, **if the system is unstable,** that is, there is at least one positive Lyapunov exponent, the bound on the error variance of the corresponding component of the state vector vanishes exponentially rapidly as the number of future observations increases. Of course, the presence of negative Lyapunov exponents means that as the number of past observations increases, the error bound goes to zero exponentially as above.

Finally, **if the system is marginally stable,** which means that there are eigenvalues on the unit circle and the system can support oscillatory solutions, the bound on the error variance for corresponding components of the state vector scales as the reciprocal of the total number of observations, both past and future. As we shall see later, the Cramér-Rao bound for chaotic systems has similar characteristics with the bound decaying exponentially rapidly for all state vector components as the number of past and future observations becomes large.

If we do not work in a coordinate system defined by the eigenvectors of the symmetric system matrix $\mathbf{G}$, the error variance of each component of $\mathbf{P}(\mathbf{x}(n))$ for fixed n and large N and M is bounded below by an exponentially small number, basically $e^{-2N\lambda_{CZ}}$, if $\lambda_{CZ} > 0$, and $e^{-2M|\lambda_{CZ}|}$, if $\lambda_{CZ} < 0$ where λ_{CZ} is the Lyapunov exponent closest to zero. This follows because each element of $\mathbf{J}^{-1}(\mathbf{x}(n))$ is a linear combination of the terms S_a^{-1}, and the largest term dominates the sum. If in fact $\lambda_{CZ} \equiv 0$, then this lower bound will be dominated by the reciprocal of the total number of observations.

Equation (11.36) offers additional insight into one's ability to estimate $\mathbf{x}(n)$ based on a finite set of noisy observations. Suppose one wishes to estimate $\mathbf{x}(n)$ at a time n near the upper or lower limits of the observations, that is for n close to $-M$ or N. For example, suppose one wishes to estimate the initial condition $\mathbf{x}(0)$ based only on the $N+1$ consecutive observations starting at time 0, i.e., $M = 0$. Then if any $\lambda_a = 0$, the bound on the error variance along the corresponding component of $\mathbf{x}(0)$ is dominated by the reciprocal of the number of observations. If all λ_a are greater than 0, and the system is unstable along all directions, the bound decays exponentially with the number of observations $N + 1$, with the decay rate dominated by the Lyapunov exponent closest to zero. If any Lyapunov exponent λ_a is less than 0, however, then

$$S_a^{-1} \to 1 - e^{2\lambda_a}, \tag{11.41}$$

and the error variance along the corresponding component of $\mathbf{x}(n)$ is bounded below by the product of the noise variance and $1 - e^{2\lambda_a}$. Note that as N increases, the element S_a^{-1} corresponding to the largest negative Lyapunov exponent λ_a dominates the diagonal in Equation (11.36) and would thus dominate the bound on the error variance of each component of the state vector $\mathbf{x}(0)$ in a rotated coordinate system. This suggests that the accuracy in estimating the value of $\mathbf{x}(0)$ is a stable system from future observations alone cannot be done with much more accuracy than the noise variance.

We have discussed this seemingly contrived example in some detail because we will argue that the same behavior transpires for a general nonlinear dynamical system where the vector field $\mathbf{F}(\mathbf{x})$ is much more interesting. However, we will show that there is a surprising twist associated with any zero-valued **global** Lyapunov exponents because of the way the local exponents approach zero.

11.4   Arbitrary, Time-Invariant, Linear Systems

For general, linear systems

$$\mathbf{x}(n+1) = \mathbf{G}\mathbf{x}(n), \qquad (11.42)$$

where $\mathbf{G}$ is not symmetric, there is no simple *analytical* expression which relates the eigenvalues of the Fisher information matrix to the Lyapunov exponents of the system.

The discussion reveals interesting properties of the eigenvalues, singular values, eigenvectors, and singular vectors of a matrix and its powers. These properties are a consequence of the multiplicative ergodic theorem restricted to linear, time-invariant systems. They do not appear to be widely known, since most people familiar with the multiplicative ergodic theorem rarely consider linear systems, and most people who deal with linear systems are not familiar with the multiplicative ergodic theorem.

The singular values of a matrix $\mathbf{G}$ are the square roots of the eigenvalues of the matrix $\mathbf{G}^T\mathbf{G}$. In general, there is no simple relation between the singular values and eigenvalues of a matrix [GV89]. Similarly there are no general relations between the eigenvectors of $\mathbf{G}^T\mathbf{G}$, the "singular vectors," and the eigenvectors of $\mathbf{G}$. Indeed the eigenvectors of $\mathbf{G}^T\mathbf{G}$ are orthogonal for nondegenerate eigenvalues, while the eigenvectors of $\mathbf{G}$ alone need not be orthogonal. An important class of exceptions consists of symmetric matrices for which the singular values are identical to the magnitudes of the eigenvalues, and the singular vectors are identical to the eigenvectors. It is because of this property of symmetric matrices, that we were able to present in the previous section a simple expression relating the Lyapunov exponents of linear systems defined by these matrices and the eigenvalues of the Fisher information matrices.

Recall how one finds the global and local Lyapunov exponents when restricted to the general, linear system given by $\mathbf{x} \rightarrow \mathbf{G} \cdot \mathbf{x}$. The global Lyapunov exponents are the logarithm of the magnitudes of the eigenvalues of

$$\lim_{L \to \infty} ((\mathbf{G}^L)^T \mathbf{G}^L)^{\frac{1}{2L}}, \tag{11.43}$$

and the local Lyapunov exponents are the logarithm of the magnitudes of the eigenvalues of

$$((\mathbf{G}^L)^T \mathbf{G}^L)^{\frac{1}{2L}}, \tag{11.44}$$

for a fixed value of L. So we see that both the global and local Lyapunov exponents are determined by the singular values of powers or iterates of $\mathbf{G}$. There are no general relations between the singular values or singular vectors of $\mathbf{G}^i$ and those of $\mathbf{G}^j$ for "small" integers i and j where $i \neq j$. This contrasts with the eigenvalues and eigenvectors of a matrix $\mathbf{G}$ and its powers: the eigenvalues of $\mathbf{G}^i$ are the ith powers of the eigenvalues of $\mathbf{G}$, and the eigenvectors of $\mathbf{G}^i$ are identical to the eigenvectors of $\mathbf{G}$.

The Fisher information matrix for $\mathbf{x} \to \mathbf{G} \cdot x$ is

$$\mathbf{J}(\mathbf{x}(n)) = \frac{1}{\nu^2} \sum_{i=-M}^{N} (\mathbf{G}^{(i-n)})^T \mathbf{G}^{(i-n)}, \tag{11.45}$$

when the noise is Gaussian, white with a variance ν^2. As this expression indicates, the Fisher information matrix is a sum of matrices defining the singular values of various powers of the matrix $\mathbf{G}$. One would not expect an analytical relation to exist between the eigenvalues of this sum of matrices and either the eigenvalues of $\mathbf{G}$ or the singular values of $\mathbf{G}$.

Despite these considerations, computer experiments [Ric94] suggest that the **asymptotic** behavior of the eigenvalues of the Fisher information matrices for general, linear systems is similar to that for symmetric linear systems. That is, the same number of eigenvalues of the inverse of the Fisher information matrix as the number of positive and zero Lyapunov exponents of the system decrease asymptotically to zero as the number of "future" observations increases, whereas the same number of eigenvalues as the number of negative Lyapunov exponents have nonzero asymptotic limits. Similarly, the same number of eigenvalues of the inverse of the Fisher information matrix as the number of negative and zero Lyapunov exponents decrease asymptotically to zero as the number of "past" observations increases, whereas the same number of eigenvalues as the number of positive Lyapunov exponents have nonzero asymptotic limits.

This "experimentally" observed asymptotic behavior of the eigenvalues of the Fisher information matrix is a consequence of Yamamoto's Theorem [Yam67], which essentially is an independently discovered version of the multiplicative ergodic theorem but restricted to linear systems. According to this theorem, the Lth roots of the singular values of $\mathbf{G}^L$ or equivalently the eigenvalues of $[(\mathbf{G}^L)^T \mathbf{G}^L]^{\frac{1}{2L}}$ converge, as $L \to \infty$, to the magnitudes of the eigenvalues of $\mathbf{G}$. By the multiplicative ergodic theorem we know that the logarithm of the magnitudes of the eigenvalues of $\mathbf{G}$ are the global Lyapunov exponents of $\mathbf{x} \rightarrow \mathbf{G} \cdot \mathbf{x}$. Thus, the logarithms of the Lth roots of the singular values of $\mathbf{G}^L$ converge to the Lyapunov exponents of $\mathbf{G}$.

As a consequence, although the singular values and vectors among smaller powers of $\mathbf{G}$ have no simple relation among themselves, for larger powers of $\mathbf{G}$, the singular vectors and roots of the singular values converge, with the logarithm of the magnitudes of the roots of the singular values converging to the global Lyapunov exponents of $\mathbf{G}$. Because of this convergence, for large enough powers of $\mathbf{G}$, the earlier discussion for symmetric linear systems is asymptotically applicable in the sense that one can perform the "same" similarity transformation on each term $(\mathbf{G}^L)^T \mathbf{G}^L$, for L bigger than some fixed value, thereby reducing each matrix to a diagonal matrix with the $2L$ powers of the eigenvalues of $\mathbf{G}$ on the diagonal. Some error terms associated with finite (i.e., small) values of the powers of the matrices accompany these eigenvalues. With this transformation, the same arguments used to describe the asymptotic behavior of the eigenvalues of the Fisher information matrix for symmetric, linear systems becomes applicable asymptotically.

In other words, for the case of a general linear system $\mathbf{x} \rightarrow \mathbf{G} \cdot \mathbf{x}$, if we wish to estimate the state $\mathbf{x}(n)$ at time n from noisy observations, the bound on the error variance for each component of $\mathbf{x}(n)$ decays approximately exponentially for any component of $\mathbf{x}(n)$ along the direction of any eigenvector of $\mathbf{G}$ for which λ_a is nonzero, as the number of past and future observations grows larger. The error variance for components along directions of eigenvectors with zero-valued Lyapunov exponents, decays only as the reciprocal of the number of observations. This is essentially a summary of the observations in this chapter.

We now derive a quantitative approximation, based on these considerations, which will prove useful in the next section. In the present model the matrix $\mathbf{J}(\mathbf{x}(n))$ has the form given in Equation (11.45), each term of which is the symmetric matrix

$$\mathbf{A}_s(i - n) = (\mathbf{G}^{i-n})^T \, \mathbf{G}^{i-n}. \tag{11.46}$$

Denote the eigenvalues of $(\mathbf{A}_s(L))^{\frac{1}{2L}}$ by $\{\exp[2\lambda_a(L)]\}$ and the eigenvectors of $\mathbf{A}_s(L)$ (and $(\mathbf{A}_s(L))^{\frac{1}{2L}}$) by $\mathbf{w}_a(L)$, so

$$\mathbf{A}_s(L)\mathbf{w}_a(L) = \mathbf{w}_a(L) \exp[2L\lambda_a(L)], \tag{11.47}$$

From the multiplicative ergodic theorem we know that as $L \rightarrow \infty$ the $|\lambda_a(L)|$ are precisely the global Lyapunov exponents of the dynamical system. Since the eigenvectors of a symmetric matrix are orthogonal and complete as a rule, we can perform a spectral decomposition on $\mathbf{A}_s(L)$ yielding

$$\mathbf{A}_s(L) = \sum_{a=1}^{d} \mathbf{w}_a(L)\mathbf{w}_a^T(L) \exp[2L\lambda_a(L)]. \tag{11.48}$$

An aspect of the eigenvalues and eigenvectors of $(\mathbf{A}_s(L))^{\frac{1}{2L}}$ suggested by numerical simulations is that as L exceeds some threshold, typically a few times ten, both the eigenvectors and eigenvalues, essentially, become independent of L. This is a consequence of the multiplicative ergodic theorem or Yamamoto's theorem, although both theorems are silent on convergence rates.

This limiting behavior makes intuitive sense. Consider the evolution of an initial state vector $\mathbf{x}(n-p)$ located p time steps before the point $\mathbf{x}(n)$ is reached, when the largest Lyapunov exponent λ_1 is positive. The system is unstable, and the directions of the sequence of vectors $\mathbf{x}(n-p)$ rapidly approach the direction of the eigenvector associated with λ_1. Because of this, the direction of $\mathbf{x}(n)$ becomes nearly independent of $\mathbf{x}(n-p)$ as p becomes large. The only vectors which do not point in the eigendirection of λ_1 are those which are the backwards iterates of the other eigendirections at $\mathbf{x}(n)$. In practice one would never see these because small numerical or round off errors grow exponentially into the expected direction. So the eigenvector $\mathbf{w}_a(i-n)$ becomes essentially independent of n as i grows large as noted above. Indeed, if one wants to find the unstable direction, i.e, the direction of greatest, long-term average growth, associated with some point $\mathbf{x}(n)$, this is one way to do it. Thus, $\mathbf{x}(n) = \mathbf{G}^p\mathbf{x}(n-p)$ points along $\mathbf{w}_1$ essentially independently of p when p is "large enough."

If we assume this to be the case, then

$$\mathbf{A}_s(L) \approx \sum_{a=1}^{d} \exp[2L\lambda_a]\mathbf{w}_a(L)\mathbf{w}_a^T(L), \qquad (11.49)$$

and the sum defining the Fisher information matrix, Equation (11.45), is identical to that for the symmetric linear system considered in the previous section, except for finite corrections:

$$\mathbf{J}(\mathbf{x}(n)) \approx \frac{1}{\nu^2} \sum_{a=1}^{d} \mathbf{w}_a(L)\mathbf{w}_a^T(L)S_a,$$

$$S_a = e^{-2n\lambda_a}\frac{e^{-2M\lambda_a} - e^{(2N+2)\lambda_a}}{1 - e^{2\lambda_a}}. \qquad (11.50)$$

Note that this is an approximation, but it represents well the behavior of $\mathbf{J}(\mathbf{x}(n))$ as M and N grow large. Since our interest is in the rate of **divergence** of the eigenvalues of the Fisher information matrix, the approximation is useful for our purposes.

With this approximation, we can approximate the Cramér-Rao bound on the error covariance matrix along directions corresponding to the (limiting) eigenbasis for $\mathbf{J}(\mathbf{x}(n))$ as follows:

$$\mathbf{w}_a^T\mathbf{P}(\hat{\mathbf{x}}(n))\mathbf{w}_b \geq \mathbf{w}_a^T\mathbf{J}^{-1}(\mathbf{x}(n))\mathbf{w}_b$$

$$\geq \delta_{ab}\nu^2(S_a)^{-1}, \qquad (11.51)$$

which, in slightly different notation, is just the result (11.36).

This approximation suggests that the general conclusions reached for symmetric linear systems also apply to arbitrary linear systems. However, remember that the inequality (11.51) is an approximation arrived at by using the multiplicative ergodic theorem to remove the L dependence of the $\lambda_a(L)$ and of the eigenvectors of $\mathbf{A}_s(L)$.

We now pose an additional, related question. What happens to estimates along the direction of a global Lyapunov exponent which is zero? For a discrete-time, dynamical system $\mathbf{x}(n+1) = \mathbf{F}(\mathbf{x}(n))$, a zero exponent would be a chance occurrence removable by varying the parameters in the vector field $\mathbf{F}$. If, however, the discrete-time dynamical system arises from a sampled differential equation rather than a Poincaré section as described earlier

$$\frac{d\mathbf{x}(t)}{dt} = \mathbf{f}(\mathbf{x}(t)), \tag{11.52}$$

then there is a **global** Lyapunov exponent which is identically zero [ER85, ABST93]. This corresponds to displacement along the direction of the flow. In fact, the direction associated with this exponent at any point $\mathbf{x}(t)$ is exactly $\mathbf{f}(\mathbf{x}(t))$.

The results for symmetric linear systems suggest that one should simply replace the value of λ_a in Equation (11.50) with zero and conclude that $S_a \approx \frac{1}{N+M}$. However, our numerical study of **local** Lyapunov exponents [ABK91] suggests that all global exponents are approached as a power of the number of steps from the initial point:

$$\lambda_a(L) \approx \lambda_a + \frac{c_a}{L^{1-q}} + \frac{c'_a}{L}, \tag{11.53}$$

where typically $0 < q \le 0.5$, and c_a and c'_a are constants associated with the particular eigenvalue. This suggests that in our approximate Cramér-Rao bound, if λ_z denotes a zero-valued global Lyapunov exponent, it should not be replaced by zero in the sum defining the Fisher information matrix, but instead by the expression $2L\lambda_z(k) = 2c_z L^q$. Now the quantity S_z still diverges for M and N large. It appears numerically [ABK91] that $c_z < 0$, that is the zero global exponent is approached from below, so it is the large M limit which is of interest. That is, information at earlier times than the observation point $\mathbf{x}(n)$ is important in the estimate $\hat{\mathbf{x}}(n)$. In this case we would conclude

$$S_z \approx \exp[2M^q|c_z|], \tag{11.54}$$

for large M. Now this is smaller than any other S_a, so in the Cramér-Rao bound it would be this term which dominates and one would conclude that the variance in one's ability to estimate $\mathbf{x}(n)$ is bounded below by $S_z^{-1} \approx \exp[-2M^q|c_z|]$. This is still zero as $M \to \infty$, and thus we still see a substantial difference between this case and that of the strictly symmetric linear time independent dynamics considered above.

11.5 Nonlinear, Chaotic Dynamics

Now we turn to the discussion of estimating the state of a system observed in noise when the system is nonlinear and has parametric settings so the orbits are chaotic. Chaos in such a system is characterized by one or more positive global Lyapunov exponents, enough negative exponents such that the sum of all exponents is negative, and, if the dynamics is governed by differential equations, one global zero exponent [ABST93]. This system we characterize by $\mathbf{x}(n+1) = \mathbf{F}(\mathbf{x}(n))$ with the understanding that discrete time dynamics comes from finite sampling of continuous time or from Poincaré sections.

First, suppose the noise is white, has zero mean, and is Gaussian as assumed earlier. Then the Fisher information matrix is

$$\mathbf{J}(\mathbf{x}(n)) = \frac{1}{\nu^2} \sum_{l=-M}^{N} [\boldsymbol{Q}^{l-n}(\mathbf{x}(n))]^T \, \boldsymbol{Q}^{l-m}(\mathbf{x}(n)), \qquad (11.55)$$

as before. The essential difference between this expression and that for the general linear result (11.45) is that the matrices entering the sum are dependent on $\mathbf{x}(n)$. This occurs because the dynamics are now nonlinear so the Jacobian matrices $\boldsymbol{Q}(\mathbf{x})$ depend on state space location. We consider now the symmetric matrix

$$\mathbf{C}_s(L, \mathbf{x}(n)) = [\boldsymbol{Q}^L(\mathbf{x}(n))]^T \, \boldsymbol{Q}^L(\mathbf{x}(n)), \qquad (11.56)$$

in which we have made explicit the dependence on the number of iterations, L, and $\mathbf{x}(n)$, the state space location we wish to estimate. Each matrix $C_s(L, \mathbf{x}(n))$ has the spectral decomposition

$$\mathbf{C}_s(L, \mathbf{x}(n)) = \sum_{a=1}^{d} \mathbf{w}_a(L, \mathbf{x}(n)) \mathbf{w}_a^T(L, \mathbf{x}(n)) \exp[2L\lambda(L, \mathbf{x}(n))]. \qquad (11.57)$$

We must consider the dependence of this matrix on both $\mathbf{x}(n)$ and L. Once again we appeal to results from the study of the **local** Lyapunov exponents which indicate that both the $\lambda_a(L, \mathbf{x}(n))$ and the eigenvectors $\mathbf{w}_a(L, \mathbf{x}(n))$ become independent of L and of $\mathbf{x}(n)$ for almost every $\mathbf{x}(n)$ in the basin of attraction of the system attractor as L becomes large. This means we can replace each of the $\mathbf{w}_a(L, \mathbf{x}(n))$ and the $\lambda_a(L, \mathbf{x}(n))$ with L and $\mathbf{x}(n)$ independent quantities $\mathbf{w}_a$ and λ_a. All conclusions made before about Cramér-Rao bounds follow once again.

There is an important aspect of nonlinear chaotic systems which is not present in any physically realizable stable, linear system. This is the fact, just noted, that the nonlinear dynamics can have both positive and negative λ_a and still have system orbits which remain bounded in state space. The boundedness comes because the full set of exponents satisfies $\sum_{a=1}^{d} \lambda_a < 0$, so volumes contract, even though some of the λ_a may be positive thus leading to everywhere unstable orbits and nonperiodic time series. The physical origin of this compactness of the system attractor is the dissipation inherent in real systems along with the folding of orbits which arises in nonlinear evolution.

The simultaneous presence of positive and negative Lyapunov exponents in a physically **stable nonlinear** system means that both the information before the observation $\mathbf{x}(n)$ we wish to estimate and the information after the observation can be utilized to establish that the Cramér-Rao lower bound on our ability to make this estimate goes exponentially to zero as the number of observations becomes large. However, this aspect of chaos—local instability, but boundedness of orbits leading to nontrivial, deterministic dynamics—comes at a price when performing state estimation. As we show in [Ric94] the boundedness of the orbits has an influence on state estimation performance, not reflected by the Cramér-Rao bound, which becomes important as the noise variance increases.

We have an heuristic argument that even when each noise term $\mathbf{v}(n)$ does not have a Gaussian distribution, the qualitative results presented thus far still apply. Suppose that the noise $\mathbf{v}(n)$ which is added to the system state vectors $\mathbf{x}(n)$ is independent and identically distributed. That is, at every time the distribution is identical and each selection of $\mathbf{v}(n)$ at any time is independent of the selection at any other time. The latter property means that the matrix (11.24) is proportional to δ_{lk}. The property of being identically distributed at all times means that the coefficient of this Kronecker delta in time is independent of time, so

$$\mathbf{S}(k,l) = \delta_{kl}\mathbf{R}, \tag{11.58}$$

where $\mathbf{R}$ is some $d \times d$ symmetric matrix. This means that the Fisher information matrix is

$$\mathbf{J}(\mathbf{x}(n)) = \sum_{k=-M}^{N} \boldsymbol{Q}^k(\mathbf{x}(n))^T \cdot \mathbf{R} \cdot \boldsymbol{Q}^k(\mathbf{x}(n)). \tag{11.59}$$

Heuristically we see that this differs from our earlier case only by the finite mixing up of directions in the matrices $\boldsymbol{Q}^k(\mathbf{x}(n))$. To make the argument we used before we need to examine the properties of the $\lambda_a(L, \mathbf{x}(n))$ and the eigenvectors $\mathbf{w}_a(L, \mathbf{x}(n))$ which enter the decomposition of the symmetric matrix

$$\boldsymbol{Q}^L(\mathbf{x}(n))^T \cdot \mathbf{R} \cdot \boldsymbol{Q}^L(\mathbf{x}(n)) = \sum_{a=1}^{d} \exp[2L\lambda(L, \mathbf{x}(n))]\mathbf{w}_a(L, \mathbf{x}(n))\mathbf{w}_a^T(L, \mathbf{x}(n)),$$

$$\tag{11.60}$$

as before. Heuristically we can view the presence of the matrix $\mathbf{R}$ as a rotation of the coordinate system at one end of the sequence of Jacobian matrices entering the composition $\mathbf{Q}^L$. Formally, though not rigorously, we can move a square root of $\mathbf{R}$ [GV89]—it is symmetric and positive definite— to each of the $\mathbf{Q}$ factors, effectively defining new $\mathbf{Q}^L$ matrices. Then as L becomes large, the argument utilized earlier, resting on the numerically observed behavior of the eigenvectors and eigenvalues of the elements of the Fisher information matrix, can be employed. While we are not prepared to present a proof of our statement, it would seem connected with the ingredients of the multiplicative ergodic theorem, so we present it here as a conjecture.

11.6   Connection with Chaotic Signal Separation

The manifold decomposition signal separation methods of Hammel and others [Ham90, FS89] rests on the ability to make a decomposition at each point of state space into linear stable and linear unstable manifolds. These manifolds are invariant in the sense that a point on them will be moved by the dynamics along them. Points on unstable manifolds map backward in time to the orbit point to which they are attached. Points on the stable manifold move forward in time to the orbit. In practice the manifolds are spanned by the set of eigenvectors corresponding to stable directions of the Jacobian matrix of the dynamics for the stable manifold and are spanned by the unstable directions for the unstable manifold. This construction fails at occasional points where stable and unstable directions coincide, so called homoclinic tangencies, but the existence of these do not change the basic idea of the method.

Once one has identified the directions in the stable and unstable manifolds, then recursively one takes the observation $\mathbf{y}(n) = \mathbf{x}(n) + \mathbf{v}(n)$ and seeks an increment, call it $\mathbf{w}(n)$, to the observation which is supposed to correct the observation and recover (that is, estimate) $\mathbf{x}(n)$ by $\mathbf{y}(n) + \mathbf{w}(n) \approx \mathbf{x}(n)$. The criterion one uses to choose $\mathbf{w}(n)$ is that the combination $\mathbf{y}(n) + \mathbf{w}(n)$ satisfy the known dynamics $\mathbf{x} \to \mathbf{F}(\mathbf{x})$. The critical numerical issue is to make sure that the increment $\mathbf{w}(n)$ starts small and remains small even though the dynamics is chaotic. This is assured by projecting the $\mathbf{w}(n)$ onto the stable and unstable directions and then mapping the projections on the stable directions forward in time and the projections on the unstable directions backward in time. For the estimation of $\mathbf{x}(n)$ information is thus used both forward and backward in time. In those cases where a stable and an unstable direction coincide the scheme fails and large increments sometime occur. In these cases the method recovers gracefully and moves back to the cleaned orbit (that is, small increments) at a rate governed by the Lyapunov exponents. The accuracy of this method was conjectured by C. Myers [Mye] to be at the rate of the smallest Lyapunov exponent, and this Chapter provides support for his conjecture. The method is applied recursively by replacing, after each pass through the data, the observed data by the present value of $\mathbf{y}(n)$ added to the increment. The process starts over with the corrected "observations" $\mathbf{y}(n) + \mathbf{w}(n)$.

When this method is applied to noisy data from the Lorenz system it works very well, and contrary to initial expectations, cleans up the component of the data along the "null Lyapunov exponent" direction. Indeed, it was thinking about that result which first led to the suggestion that local Lyapunov exponents had a role to play in signal separation.

Using this stable-unstable decomposition method and full knowledge of the dynamical system, it is possible to demonstrate remarkable ability to recover the clean chaotic signal when the original signal to noise level is about 10 to 15 dB or more. Beyond that the linearity used when the increment is small begins to fail. Our own interest here stops at this point. We are not so much concerned with any specific method of signal separation or state estimation but more interested in establishing the connection between the arguments in this chapter and demonstrated methods which appear to have the power to achieve the best performance the Cramér-Rao bound would allow. It is quite plausible to us that other methods could achieve these limits, but not knowing them, we cannot comment further.

11.7   Conclusions

In this chapter we have examined the ways in which estimating the state of a nonlinear, chaotic system given noisy observations, differs substantially from familiar considerations for linear systems [Tre68]. The Cramér-Rao bound is the only one considered here (see [Ric94] for studies of other bounds on estimation and the role of nonlinearity for them), and we have seen that the properties of the Fisher information matrix, which provides the lower bound for the variance of the estimation of the true state from noisy observations, are intimately connected with the Oseledec matrix which enters the multiplicative ergodic theorem [Ose68] and with the determination of local and global Lyapunov exponents for the system which governs the state of the system.

We found that if one makes observations in discrete time "n" from the far past $n = -M$ into the far future $n = N$ $(M, N > 0)$, then the variance in the error associated with the estimate of $\mathbf{x}(n)$ at some finite time "n" decreases exponentially rapidly with the number of observations used in the estimate. We associated this in an heuristic geometric way with the overlap of a sphere of noise surrounding an orbit point $\mathbf{x}(n)$ when those noise spheres have been moved forward and backward in time according to the nonlinear dynamics. The stable directions contained by the sphere will shrink exponentially fast when moved forward, and the unstable directions will shrink exponentially fast when moved backward. The overlap defines where the estimate of $\mathbf{x}(n)$ must lie, and thus the ability to estimate exponentially accurately takes on less mystery.

The only additional point of some interest is the role played by the local Lyapunov exponent which corresponds to zero global Lyapunov exponent. One's first thought would be that this zero exponent would not allow for shrinking of noise spheres in either the forward or the backward directions. However, the local Lyapunov exponent only approaches zero as a power of the number of steps along a trajectory which one has moved, and it is this power which determines the smallness of the error estimate one can achieve. The quantitative argument for this has been given above.

We anticipate that the analysis of other error estimates and the bounds associated with them will suggest further ideas for actually achieving the performance that they imply in estimating system states in the presence of contamination.

12

Summary and Conclusions

The theme of this book has been the development of tools for the analysis of observed nonlinear systems. Primarily we have focused on ways to analyze chaotic motions of nonlinear systems as such motions are continuous, broadband spectrally and clearly require different tools for their analysis than linear systems operating in modes producing narrowband signals. Along the way we have pointed out places where the time domain, state space methods of analysis, especially those which are local in multivariate state space, have interesting applications to issues little connected to chaos. The modeling of inputs in terms of output phase space and the representation of one dynamical variable in terms of the reconstructed state space of another, are two examples of this.

Along the way we have digressed to touch on other topics of interest to the author and connected, at least in my own mind, to the main theme. The bound on one's ability to estimate a signal in chaos treated in Chapter 11 is one of these, but control and synchronization of chaos also fall into this category.

It seems quite important as one develops and uses various methods for analysis of chaotic observations that one try to keep in mind the goals of that analysis. There is no general statement of these goals; they depend on the individual and the problem at hand. As interesting as it is at times to work on general tools to choose a time delay for phase space reconstruction or new ways to establish whether one has unfolded an attractor, these cannot be the end of the line in developing the analysis tools. The tools do not have an interesting life, for a physicist anyway, outside of physical problems to which they might apply to allow one to uncover aspects of those problems which remain hidden by other methods. The same could well be said for the finest of the tools for linear analysis: the fast Fourier transform. Outside of its use in extracting the spectral content of signals where spectral content is important, it remains only an interesting but not very special device of applied mathematics. Once it is brought to bear on learning something about real dynamical systems in Physics or Biology or Engineering or other fields, it acquires a substantial value. This is a rather pragmatic point of view, of course, and while it does not deny the genuine academic pleasure of exploring whatever comes along, it does eschew trying anything and everything just because it is there. Not everyone will agree that all topics in this book strictly adhere to my own admonition, but there has been some attempt to toe the line.

The problem with developing analysis tools in the absence of something to analyze, that is sensibly collected data from a system of intrinsic interest, is that it can quickly become sterile. The chapter on examples from actual laboratory field experiments was designed to move away from the starkness of exposing method after method to demonstrate how one might learn things of some interest using the methods at hand.

This book has not, by any stretch of the imagination, attempted to be a review of all methods developed for the analysis of chaotic data. Frankly I am not sure that such an effort would be more than a compilation of the essential parts of hundreds of excellent research papers published over the past 15 years, and in that sense it might be encyclopedic but not very helpful. This book has unabashedly been a focus on what I think has been useful in exploring the Physics of nonlinear dynamical systems. I am sure that other methods are equally useful, answer questions just as well, and sometime answer questions beyond the scope of this book. I am equally certain that the methods explained in this book will be useful far beyond the bounds of Physics alone. Indeed, I believe that in the analysis of biological systems, at least small systems perhaps less complex than, say, the brain as a whole, these methods prove quite helpful. As a physicist, I have felt quite comfortable not trying to pretend to a knowledge of biology or other topics beyond fact, and selected my examples accordingly. This may have excluded biologists and others from the reading audience, but I trust they will see the earlier chapters as addressed to them as well. It often comes as a surprise that in the 1990s everyone does not have a familiarity with the Heisenberg equations of quantum mechanics, but that is apparently so.

In the remainder of this closing chapter we want to give a "cloudy crystal ball" view–clearly personal–of what has been accomplished in the study of chaotic motions using data from observing them and what may be the subject of further study in the coming years. This is strictly speculative, so the reader may wish to simply pass on this and return to using the parts of this monograph which seem to work.

12.1   The Toolkit–Present and Future

The study of nonlinear dynamics has taken essentially one of two roads for much of its past:

i starting from dynamical equations, differential or discrete time, of interest chosen for whatever reasons, the bifurcation sequence of fixed points, periodic orbits, etc has been analyzed analytically where possible and, more recently, numerically when required. The collection of the possible behaviors of the particular nonlinear system and implications for other nonlinear systems in the same "class" may constitute the main product of the investigation. If the equations have a real experimental or engineering context, then the bifurcation diagram may serve to support the choice of the equations and possibly assist in determining unknown parameters in the original equations. The equations would then be available for further use in engineering design or other applications, if any.

Many studies of nonlinear equations have been for the purpose of investigating what kind of behavior is possible at all in nonlinear systems. This reverses a long standing myopic demand for analytic solutions as those of any real scientific interest, and in this regard this has been and will continue to be a quit important enterprise.

While we spent some time in the early chapters of this monograph on various features of this classification of phenomena aspect of nonlinear dynamics, we have contributed here nothing to that direction of study. It is quite important to have that kind of analysis as part of the background for the main topic of this book, as a context is required to interpret dynamics when its consequences are observed.

ii starting from observed data of interest for whatever reasons an analysis methodology has been created which is intended to illuminate properties of the source of the data. This is often called the "inverse" problem of nonlinear dynamics, especially by persons who have concentrated on the issues just above, and it constitutes the main thrust of this book. The goal is to use this analysis to determine equations of motion—flows or maps–for the source which are consistent with the data. "Consistent with the data" takes on a different purport, as we have stressed repeatedly, than the traditional comparison of observed orbits in phase space. As we know one cannot do this in chaotic motions because every orbit is unstable. Consistency then is established in terms of comparison of invariants of the dynamical source of the data, and the tools for doing this have been discussed at length here.

The "inverse" problem may become indistinguishable at this point from the "direct" problem above, if one has global phase space evolution rules and can connect them to familiar dynamical variables in the problem. These could be pressure, voltage, velocity, etc. The inverse problem may remain quite different if the solution has been couched in terms of local dynamical rules operating only on or near the strange attractor.

Broadly speaking the separation between these two paths is quite small. Both have constituted the concrete realization of how we learn about interesting dynamics. The only new aspect injected by recent studies has been the acceptance that in analyzing nonlinear systems concrete answers may only be available numerically, and models which are somehow little more than well constructed lookup tables may actually constitute the best answer one can achieve. In my mind this does not set aside the attempt to find more traditional models which, at their core are expressions of $F = ma$ in one way or another, but allows an understanding to develop which is more geometric and perhaps more qualitative than found in traditional physics texts.

The analysis tools found in this book are in many ways an expansion of Table 1.1 which is repeated on the next page for pedagogical reasons. The essential recognition [ER85] that we could learn the heart of the nonlinear dynamics of the source of some observations by using the information contained in a single time series of scalar measurements stands at the foundation of all we have done in this book in trying to fill out the demands of this table.

We have not wandered afield from the use of time delay embedding as the means of reconstructing multivariate phase space from scalar observations. This is not meant to be narrow minded as other ways are well explored [MSN*91]. As a general rule which has wide and often simple applicability, the use of time delays of the observations $s(t_0 + n\tau_s) = s(n)$ to form

$$\mathbf{y}(n) = [s(n), s(n + T), \dots, s(n + (d_E - 1)T)] \qquad (12.1)$$

is to be highly recommended. Once this is agreed on, then we can entertain various ways to choose the time delay T and the global embedding dimension d_E. Indeed, if one does not have data which is equally sampled in time—no single τ_s—it may not be the best choice. A form of global data vector with differing time delays in each component may be wise.

While I have stressed the nonlinear correlation properties among the $s(n)$ as a way of choosing T and attempted to bring them out by the use of the average mutual information among the measurements constituting the components of $\mathbf{y}(n)$, this is by no means dictated by the general phase space reconstruction idea. Among the various tools discussed in this book, I would be unsurprised to see another form of "nonlinear autocorrelation" among the observations replace the average mutual information methods. This is not to disparage mutual information as a meritorious measure of properties of a nonlinear source of signals, but as important as mutual information has proven in communications [Gal68], the factorization of a

Linear signal processing	Nonlinear signal processing
Finding the signal **Signal separation** Separate broadband noise from narrowband signal using spectral characteristics. System known: make matched filter in frequency domain.	*Finding the signal* **Signal separation** Separate broadband signal from broadband "noise" using deterministic nature of signal. System known: use manifold decomposition. Separate two signals using statistics on attractor.
Finding the space **Fourier transforms** Use Fourier space methods to turn differential equations or recursion relations into algebraic forms. $x(n)$ is observed; $x(f) = \Sigma x(n) \exp[i2\pi n f]$ is used.	*Finding the space* **Phase space reconstruction** Time lagged variables form coordinates for a phase space in d_E dimensions: $$\mathbf{y}(n) = [x(n), x(n+T), \ldots,$$ $$x(n + (d_E - 1)T)]$$ d_E and time lag T using mutual information and false nearest neighbors.
Classify the signal Sharp spectral peaks. Resonant frequencies of the system *Quantities independent* *of initial conditions*	*Classify the signal* Invariants of orbits. Lyapunov exponents; various fractal dimensions; linking numbers of unstable periodic orbits *Quantities independent* *of initial conditions*
Make models, predict $x(n+1) = \Sigma c_j x(n-j)$ Find parameters c_j consistent with invariant classifiers—location of spectral peaks.	*Make models, predict* $$\mathbf{y}(n) \to \mathbf{y}(n+1)$$ as time evolution $$\mathbf{y}(n+1) = F\mathbf{y}(n), a_1, a_2, \ldots, a_p]$$ Find parameters a_j consistent with invariant classifiers—Lyapunov exponents, fractal dimensions. Models are in local dynamical dimensions d_L; from local false nearest neighbors. Local or global models.

TABLE 12.1. Comparison of Linear and Nonlinear Signal Processing

joint probability density as a measure of independence of two measurements is the central theme, and it may well be that other theoretical constructs will prove of some use in practical algorithms.

Further, since the underlying embedding theorem [Man81, Tak81, CSY91] makes no reference to the value of T except that it should not be exactly a low order integer multiple of the period of a periodic orbit, it is hard to see how a compelling argument could be made for or against any prescription for choosing the time delay T in the vectors $\mathbf{y}(n)$. It may well depend on the question one wishes to ask about the data and the uses to which one wishes to put the answers which would dictate the choice of prescription. A sage piece of advice is to explore several choices and think about the varying answers, if indeed they vary, in the context of the specifics of the signal source.

The choice of global embedding dimension d_E should be based on the concepts in the basic theorem which one is trying to implement in algorithmic form. The false nearest neighbors method [KBA92] certainly tries to do that by its focus on the idea of identifying phase space neighbors which are created by projection of orbit strands from distant parts of an attractor. By adding additional coordinates until these nondynamical neighbors are removed, we directly address the matter of the number of dimensions d_E one needs to satisfy the key geometric idea in the theorem.

False nearest neighbors is a pretty good algorithm. Recognizing its problems when data is oversampled and when the spectrum of the data is very steep so some degrees of freedom with very small numerical effect are not properly emphasized, one can systematically make modifications which iron out the difficulties. There is no doubt that false nearest neighbors will be improved as time goes by, but since it goes to the heart of the embedding theorem [CSY91] many features will remain.

The true vector field method [KG92] also heads in the right direction, but attempting to construct local vector fields suffers from having to take differences between data points, and this can lead to numerical noise as difference taking is a high pass filtering operation. To date the true vector field method has not been tested as extensively on empirical data, and it may prove just as valuable. Again the sage advice is to try both as they seek information about the same basic aspect of the attractor and when faced with realistic data, one can often use all the help available.

The next step in our list is that of classifying the dynamical system with invariants insensitive to initial conditions. This is a woefully bare region. We know about Lyapunov exponents and fractal dimensions and for three-dimensional dynamics something about topological invariants. There is no sense of completeness or being able to exhaust the information required to classify strange attractors, and this is a gap which mathematical insight would do well to close. If we had further invariants, both of the dynamics and of the coordinate systems in which we view the dynamics, we could with some confidence identify this or that attractor and, perhaps, the dynamical source of the signal. This would be the nonlinear equivalent, I guess, of deducing the shape of a drum from its acoustic tones.

An area rather untouched by mathematics in the study of dynamical systems is that of local properties on an attractor. These tend to be quite system specific, and it, understandably, has been global properties which have drawn the attention of rigorous analysts. Our look at local or finite time Lyapunov exponents revealed that they are what is important for prediction, since one cannot predict in the long run when the orbits are chaotic. The various moments of these might help in the classification issue. Another direction would be to identify the analogues of Lyapunov exponents for the nonlinear parts of the stable and unstable manifolds which are associated with orbits in phase space. It is only the properties of the linear manifold which are touched on by Oseledec' Multiplicative Ergodic Theorem. The numerical issues associated with these nonlinear invariant manifolds are difficult, but perhaps a theorem or two of the Oseledec' variety yielding attractive goals would bring forth good algorithms.

Signal separation is a very interesting task with implications far beyond just "noise" reduction, though that too can be valuable. Identifying properties of chaotic signals, using their phase space structure and their deterministic features, will certainly allow the removal of contamination with different structure and deterministic rules. The idea that one could systematically examine long time series of "noisy data", with an eye toward identifying those data which have low dimensional structure, seems quite attractive. This knowledge would then be useful in separating this kind of contamination when it is encountered in conjunction with either a regular signal of interest or for that matter a chaotic signal of interest. The various methods outlined in our chapter on this subject can barely have touched the surface on what one should be able to do to take advantage of phase space structure when it is found, and I anticipate that what we now know in that area especially will be supplanted by significant developments in the near future. The untapped opportunities are surely substantial.

Direct noise reduction, when knowledge of the signal against which one wishes to reduce the noise or knowledge of the statistics of the high dimensional noise are absent, may end up being one or another strategy for filtering and averaging in phase space. This may parallel the developments for linear systems and evolve to a series of techniques, quite useful ones probably, which depend on one's assumptions about the statistics of the contamination. What could be a helpful development is a way to utilize some general class of dynamical properties of the attractor structure in phase space to assist in the separation of noise from a signal of interest. This is open territory.

12.2   Making 'Physics' out of Chaos–Present and Future

Actually doing something with the results of all of one's fine analysis is the proof of the pudding. This includes prediction, foremost, and then control, synchronization, and other applications as byproducts. Some of these we have alluded to or developed in this book, but it is likely that much more is ripe for creation within the context of specific real world and laboratory problems. Prediction, as we saw, is in many ways quite simple, and depending on the demands one places on it, impossibly hard. There is no question that making models in phase space, which take advantage of the local structure of strange attractors, works well. Local polynomials are the least clever way of utilizing this structure, and one may be certain that the use of richer classes of basis functions will prove very widespread and very useful.

My own view may pass from the scene, but I accept the idea that familiarity with quantum mechanics or fluid dynamics in a traditional analytic mode, as limited as we recognize this to be for nonlinear dynamics, is critical for moving away from the useful but mechanism neutral world of local models in phase space. Given Rissanen's [Ris89] reminder of the idea of Kolmogorov that the number of models one could devise is so large that we really have no chance of algorithmically deducing the structure of dynamics, I see no substitute on the horizon for choosing among dynamical forms which have worked whether they are classical or quantum mechanical or some thoughtful blend [GH93].

From the point of view of the physicist or biologist or natural scientist in general who wishes to use the methods discussed in this book, I recommend a very healthy skepticism about being impressed with the success of local predictive models. One can view that success as some evidence that the structure and characteristics of the attractor corresponding to the observed behavior is correct, and this will then lead to a framework, and that is all, for examining the equations of motion deduced from other information and experience.

There are important qualitative features one can abstract from the existence of chaotic behavior in observed systems and the demonstration, in a quantitative sense, of that chaos using the methods we have presented. For example, in some small neural systems from invertebrates, it appears that the collective behavior of a dozen or so neurons to perform gastric functions exhibits low dimensional chaotic behavior which is predictable in a manner consistent with the Lyapunov exponents deduced from the data [AHR*95]. Despite the absence of the details of any mechanism for this effect, one may well ask what is the function of the chaos in such a system. Perhaps the answer is similar to the answer one gives when analyzing the value of control theory in the presence of chaos: when a nonlinear system can be chaotic, it opens up an enormously larger domain of phase space for the system motions to explore. In the case of control, more options are thus opened up to which one may direct the behavior, as desired. In the case of biological chaos, perhaps the options presented for dynamical diversity lie at the heart of survival in the face of changing environmental settings. Absent nonlinear behavior and the possibility of chaos, the ability of the biological system to alter or adapt or evolve would be so limited it would find no success in persisting. All this is the kind of worthwhile, broad speculation engendered by the ability to extract quantitative substance from seemingly complex time dependence, and to this extent it opens up innovative views of that time dependence it has served well. The detailed, quantitative description of that dynamics will most likely go back to expressions of forces and responses within the context of various forms of mechanics.

12.3   Topics for the Next Edition

This could easily be the longest section of this monograph and could comprise another by itself. Three items which will surely be of great interest are these:

i **Tools for the analysis of dynamics in time/space.** Distributed systems or field theory represent a large jump over present workable methods in the challenges they present conceptually and computationally. We have concentrated our focus in this book on measurements at a single spatial point or observations of "lumped" dynamical systems where averages over detailed spatial behavior are done automatically in the experimental setup. Our examples of the chaotic laser, the turbulent boundary layer, the nonlinear low frequency circuits, and the Great Salt Lake volume all fall into this category. Indeed, much of the reason for the success of the toolkit developed here (and summarized by Table 1.1) has been this removal of spatial degrees of freedom.

The introduction of spatial independent variables, mathematically equivalent to an infinite number of degrees, certainly complicates matters. Interest in chaotic fields is driven, as it should be, by both academic concerns and by the practical fact that many phenomena are described by fields, not by the simplification of a few ordinary differential equations. The realistic convection problem addressed by Salzmann [Sal62] and Lorenz [Lor63] over 30 years ago which led to the blossoming of our knowledge in chaos remains, firmly set in the dynamics of atmospheric fields.

There has been a substantial amount of work in this area with much of the accomplishment focused on the behavior of systems just after bifurcation from steady or periodic states [CH93, LM94] where issues of pattern formation outside of equilibrium are addressed. There has also been a lot of analysis of spatio-temporal chaos which has taught us a great deal about the range of phenomena associated with fields. This work has been along the lines of the first path suggested above where definite equations are suggested by experiment or by general arguments or by a sense of their mathematical generality, and these equations are analyzed analytically and numerically to extract their universal or generalizable behavior, where it is possible.

Efforts along the lines of the theme we have addressed are less well developed, and it was the opinion of this author that a chapter or two on the subject of space/time dynamics was not yet appropriate. This is one of the richest areas for this kind of work to concentrate on.

ii **Classical/quantum chaos connections** Chaos in nonrelativistic quantum mechanics is mathematically not possible when the spectrum of states is discrete. The argument is simple. When the Hamiltonian operator has a discrete spectrum $E_1, E_2, ...$, the time dependence of any Heisenberg operator $\mathbf{A}(t)$ is composed of only the frequencies $E_i/\hbar$. The general matrix element $< a|\mathbf{A}(t)|b >$ develops in time as

$$< a|\mathbf{A}(t)|b > = < a|e^{\frac{iHt}{\hbar}}\mathbf{A}(0)e^{\frac{-iHt}{\hbar}}|b >$$
$$= \sum_{m,n} < a|m > < n|b > < m|\mathbf{A}(0)|n > e^{\frac{i(E_m - E_n)t}{\hbar}}, \qquad (12.2)$$

and the Fourier spectrum is that of quasi-periodic, not chaotic motions.

When one considers other aspects of quantum predictions such as the energy spectrum of quasi-classical states or scattering states, the properties of classically chaotic systems do make their appearance and provide interesting phenomena [Ott93]. The detailed connection between the clearly quantum mechanical states where one has a few quanta, and semi-classical states where one has many quanta remains a fruitful area of investigation. In the example we described of chaotic laser intensity fluctuations, one has many, many photons, so it falls clearly into the semi-classical category. Our successful analysis of that data sheds no light on the connection between few quanta and many quanta dynamics. One would hope that experiments can be devised which would bridge this connection and illuminate for us how it is that the required coarse graining to average over small scale quantum fluctuations to produce classical motions [GH93] is carried out by our usual "worked example"—natural phenomena. It is not at all clear to me how the tools developed in this book apply to this class of issues; perhaps they do not.

iii **Practical applications** As with the development of many fundamental areas in natural sciences there is an implicit goal of finding new practical, perhaps commercializable, applications for the discoveries. As with other areas the use of new ideas in chaos and nonlinear dynamics takes time as the generalizable and utilizable lessons are absorbed by the practitioners of existing art. At the time of completion of this book (Winter, 1995) the detailed technical applications of this area of new knowledge remain few. Utility in communications, predictability of previously designated "noise", perhaps applications in mechanics, stand out as arenas where these ideas have found their earliest applicability.

It seems to me quite an obvious proposition that since physical systems can behave chaotically, and many do, that there is a wide range of fascinating applications of these physical behaviors just as there has always been for every other class of new physical phenomena. The movement of the ideas in this book and others of similar orientation [Moo92, Ott93] from the laboratory to engineering design tools and to the assembly line must be a matter of time and education of a generation of new practitioners. The prospect is most attractive.

Appendix

A.1 Information Theory and Nonlinear Systems

We will be concerned with **dynamical systems whose orbits are unstable everywhere in phase space**. This does not mean that any initial point $\mathbf{x}(0)$ goes to infinity in time, only that points nearby at some time move rapidly away from each other at later times. If we have two points $\mathbf{x}_A$ and $\mathbf{x}_B$ in the phase space of a nonlinear dynamical system and their distance at t_0 is $|\mathbf{x}_A(t_0) - \mathbf{x}_B(t_0)|$, then typically their distance at some $t > t_0$ is

$$|\mathbf{x}_A(t) - \mathbf{x}_B(t)| \approx |\mathbf{x}_A(t_0) - \mathbf{x}_b(t_0)|e^{\lambda(t-t_0)} \quad \lambda > 0. \tag{A.1}$$

This cannot hold for all $t > t_0$ or points would run off to infinity and the dissipation prevents that. It does hold for some time, basically until dissipation can make its presence manifest, then other global properties of the dynamics set in. The value of λ depends in detail on the parameters entering the vector field $\mathbf{F}(\mathbf{x})$, and any value of $\lambda > 0$ gives rise to what we call chaos.

$\lambda > 0$ cannot occur for any global linear dynamics which makes physical sense. A linear system is of the form

$$\frac{d\mathbf{x}(t)}{dt} = \mathbf{A} \cdot \mathbf{x}(t), \tag{A.2}$$

and the dynamics is entirely governed by the eigenvalues of the matrix $\mathbf{A}$. If any eigenvalue has a positive real part, the orbits $\mathbf{x}(t)$ go to infinity and the dynamical description by linear equations is wrong. Global linear dynamics requires that the real part of any eigenvalue λ of the matrix $\mathbf{A}$ be negative or zero.

In nonlinear dynamics this is not the case. One may have, and routinely does have, $\lambda > 0$ for any starting point in the phase space, and still because of the global dissipation, orbits remain bounded. The implication of $\lambda > 0$ is that orbits are unstable everywhere and never close on themselves, leading to **nonperiodic** motions $\mathbf{x}(t)$.

Another implication is that the instability reveals information about the system which is not uncovered in regular linear dynamics. We imagine that in our observations of the dynamics we have a fixed size error sphere in the phase space. This error sphere is a property of the observation instruments, not a property of the dynamics. If two phase space points $\mathbf{x}_1$ and $\mathbf{x}_2$ lie within this error sphere at some time t, we cannot know this by our observations as they are within experimental error. At a later time $t' > t$, these points will have moved apart because of the instabilities everywhere in phase space. The distance at t' is related to the distance $|\mathbf{x}_1 - \mathbf{x}_2|$ at t by

$$|\mathbf{x}_1 - \mathbf{x}_2| e^{\lambda(t'-t)} \quad \lambda > 0, \qquad (A.3)$$

and this grows rapidly. Soon it is larger than the error ball, so one can now experimentally distinguish the new locations of the two points which had earlier been hidden in the errors. This uncovering of information is specific to nonlinear systems with instabilities. These are precisely the conditions which lead to chaos, so we anticipate the information will be a critical idea in the discussion of chaotic, namely unstable, irregular, nonperiodic motions on strange attractors.

The quantification of information takes the form of the answer to the question: if we make a series of measurements $a_1, a_2, \ldots, a_m, \ldots, a_M$ and another series of measurements $b_1, b_2, \ldots, b_n, \ldots, b_M$, how much do we learn about the measurement a_m from the measurement b_n ? In *bits* the answer is [Gal68]

$$\log_2 \left[\frac{P_{AB}(a_m, b_n)}{P_A(a_m) P_B(b_n)} \right], \qquad (A.4)$$

where $P_A(a_m)$ is the normalized histogram of the distribution of observed values in the A measurements a_m. $P_B(b_n)$ is the same for the b_n observations, and $P_{AB}(a_m, b_n)$ is the joint distribution for both measurements. If the measurements are independent, then as a general nonlinear statement this implies

$$P_{AB}(a_m, b_n) = P_A(a_m) P_B(b_n), \qquad (A.5)$$

and the amount learned about one measurement by making the other is zero as it should be. The logarithm of this ratio of normalized distributions, probabilities if you will, is called the **mutual information** between the two measurements.

Mutual information is no different from other variables defined on the space of measurements (a_m, b_n). It is distributed over the space of measurements (a_m, b_n), and its moments have the usual meaning. Its average over all measurements or equivalently its mean value, called the **average mutual information**, is

$$I_{AB} = \sum_{a_m, b_n} P_{AB}(a_m, b_n) \log_2 \left[\frac{P_{AB}(a_m, b_n)}{P_A(a_m) P_B(b_n)} \right]. \qquad (A.6)$$

This is a kind of nonlinear "correlation" function between the two sets of measurements as we now shall see.

Suppose the two measurements were some voltage $v(n); n = 1, 2, \ldots$ as the a_m measurements and the *same* voltage a time lag T later $v(n+T); n = 1, 2, \ldots$ for the b_n measurements. The average mutual information is a function of this time lag and acts as the **nonlinear autocorrelation function** telling how, in a nonlinear way, the measurements at different times are connected on the average over all measurements.

This nonlinear autocorrelation function allows us a sense of when the measurements of voltage $v(n)$ and voltage $v(n + T)$ are independent. At $T = 0$ these are clearly not independent as they are the same. The value of $I_{v(n), v(N+T)} = I(T)$ at $T = 0$ happens to be the entropy as conventionally defined for a distributed quantity [Gal68]

$$I(0) = - \sum_{a_m} P_A(a_m) \log_2 [P_A(a_m)]. \qquad (A.7)$$

This is the amount of information we can gather about the source of measurements or perhaps should be called the capacity of the source to generate information. Through a theorem of Pesin [Pes77, ER85] it is related to the indices of stability λ discussed above.

When the system is unstable we anticipate that for large T the measurements $v(n)$ and $v(n+T)$ have little to do with each other. Clearly they are independent. When we discuss creating multidimensional representations of the observed data from observations of a single voltage, we shall accept a prescription for independent but not disjoint observations, and we shall state it in terms of the nonlinear autocorrelation $I(T)$.

A.2   Stability and Instability

Chaos can be characterized as a state of a nonlinear oscillator which is unstable everywhere in its phase space. This is a rather remarkable kind of state and certainly represents a change in worldview from what we are directed to consider in our mainline education. This state of nonequilibrium motion arises when a dissipative or conservative (Hamiltonian) system is stressed by external forces and regular motions and becomes less efficient in transporting energy, momentum, etc. through the system [Cha61]. The study of the sequence of instabilities encountered as one varies the stress is known as bifurcation theory.

To approach this new view of allowed system state, we will start on familiar ground, namely, with a fixed point or time independent equilibrium state, and consider how stressing a system can lead to chaos. It is not our intention to replace the extensive literature on bifurcation theory [Cra91, Dra92, GS88]. Not only would that distract from the main goal of this monograph, but there is no doubt we could not do justice to the subject. Fortunately there are excellent elementary and advanced texts on this subject as well as many very well done review articles covering more complex and subtle aspects of the subject.

A.2.1 Lorenz Model

We are interested in a qualitative view of sequences of instabilities so we will focus our attention on a specific, important example. This is the set of three differential equations of Lorenz [Lor63] abstracted from earlier work by Salzmann [Sal62] on convection of the lower atmosphere. The equations are deduced from the infinite dimensional partial differential equations governing momentum transfer, namely, the Navier-Stokes equations, and heat transfer. The equations for the fields of velocity and temperature are expanded in spatial basis functions which meet the required boundary conditions and the evolution equations for the time dependent coefficients are then analyzed. This is called a Galerkin approximation, and if the stress on the system is large, there may be a very large number of ordinary differential equations required to accurately describe the phenomena of interest. The number of relevant coefficients may well depend on the basis functions chosen.

In the Lorenz approximation to these equations only three coefficients were kept. One for the velocity field and two for the temperature field. There are three parameters which reflect the key physics of the situation. The Rayleigh number r is a dimensionless measure of the temperature difference between the top of the atmosphere and the ground. This represents the stress on the atmosphere giving the instability. It leads to convection and fluid and thermal transport from ground, heated by sunlight, to the top of the atmosphere where infrared radiation departs, cooling the system. σ is the Prandtl number which measures the ratio of momentum transport by viscosity and heat transport by thermal conductivity. The third b is a measure of length scales in the convective process.

The conductive state, with zero fluid velocity and fluid temperature varying linearly in the spatial coordinate along which gravity and the temperature gradient lie, is removed from the fields. The resulting, rescaled equations for the Galerkin coefficients are remarkably simple in appearance:

$$\frac{dx(t)}{dt} = \sigma(y(t) - x(t)),$$

$$\frac{dy(t)}{dt} = -x(t)z(t) + rx(t) - y(t),$$ (A.8)

$$\frac{dz(t)}{dt} = x(t)y(t) - bz(t).$$

The first question one asks of such equations is the location of its fixed points in the dynamical variables $\mathbf{x}(t) = [x(t), y(t), z(t)]$. Fixed points are time independent solutions of the equations. This means the righthand side of Equation (A.8), called the vector field $\mathbf{F}(\mathbf{x})$, is zero. The zeros are at the locations

$$[x, y, z] = [0, 0, 0],$$ (A.9)

and

$$[x, y, z] = [\pm\sqrt{b(r-1)}, \pm\sqrt{b(r-1)}, r-1].$$ (A.10)

The state $[0, 0, 0]$ is pure thermal conduction of the heat applied at the lower boundary and carried to the top boundary. This state is stable to linear perturbations when the equations linearized about it have solutions which relax back to it from any initial small perturbation. The equations linearized about $[0, 0, 0]$ are

$$\begin{pmatrix} \dot{x}(t) \\ \dot{y}(t) \\ \dot{z}(t) \end{pmatrix} = \begin{pmatrix} -\sigma & \sigma & 0 \\ r & -1 & 0 \\ 0 & 0 & -b \end{pmatrix} \begin{pmatrix} x(t) \\ y(t) \\ z(t) \end{pmatrix}.$$ (A.11)

The stability of these deviations from $[0, 0, 0]$ are governed by the eigenvalues of the matrix in Equation (A.11). These eigenvalues are $\lambda = -b$ and the solutions to

$$\lambda^2 + (\sigma + 1)\lambda - \sigma(r - 1) = 0.$$ (A.12)

If $r < 1$, all eigenvalues are negative. At $r = 1$, one of the eigenvalues is zero, and the other is $\lambda = -(\sigma + 1)$. For $r > 1$, there is an eigenvalue with a positive real part.

The solutions to the linearized differential equations are

$$\mathbf{x}(t) = \sum_\lambda e^{\lambda t} \hat{e}_\lambda x_\lambda,$$ (A.13)

where the unit vectors $\hat{e}_\lambda$ are along the eigendirections of the matrix above, and the values x_λ are the projections of the initial perturbation $\mathbf{x}(0)$ along these directions at $t = 0$. If all of the λ have negative real parts, as is the case when $r < 1$, then all perturbations tend to zero, and we conclude the state $[0, 0, 0]$ is stable to small perturbations. For $r > 1$, the conduction state $[0, 0, 0]$ is linearly unstable and a new state of the system is reached under the nonlinear evolution.

Physically what is happening here corresponds to increasing the temperature difference between the bottom and the top of the atmosphere, that is, increasing the value of r. When r is small, the fluid dynamical system simply conducts heat from the hotter bottom to the cooler top. This process is stable as long as the temperature difference is small enough, that is $r \leq 1$, but becomes unstable when r becomes greater than unity. At that point the system prefers to move into another state which transfers the heat more efficiently and in a different manner. Conduction involves no macroscopic fluid motion; indeed, it is simply diffusion of heat. When $r > 1$ this becomes the less preferred mode of heat transfer, the fluid begins to move and heat is transferred not only by conduction but also by bodily transport of warm molecules from lower to higher in the atmosphere. This is convection, and is represented by the other fixed points

$$[\pm\sqrt{b(r-1)}, \pm\sqrt{b(r-1)}, r-1], \qquad (A.14)$$

where the signs refer to opposite directions of motion of the fluid. In this state warm fluid rises in certain regions and about a distance b away cooler liquid falls. The basic physical mechanism here is tied to the equation of state and to buoyancy in the gravitational field. Warm fluid is lighter, so it tries to rise. If the temperature difference is too small, $r < 1$, the buoyancy force cannot overcome friction, but when $r \geq 1$, the opposite is true.

So for $r > 1$, we look at the other linearized fixed points of the Lorenz vector field. The equations of motion about these point are

$$\begin{pmatrix} \dot{x}(t) \\ \dot{y}(t) \\ \dot{z}(t) \end{pmatrix} = \begin{pmatrix} -\sigma & \sigma & 0 \\ 1 & -1 & x_\pm \\ -x_\pm & -x_\pm & -b \end{pmatrix} \begin{pmatrix} x(t) \\ y(t) \\ z(t) \end{pmatrix}, \qquad (A.15)$$

where $x_\pm = \pm\sqrt{b(r-1)}$. The eigenvalues of this matrix all have negative real parts until r is equal

$$r_c = \frac{\sigma(\sigma + b + 3)}{(\sigma + b - 1)}, \qquad (A.16)$$

at which value $x_\pm$ also become unstable. For $1 < r < r_c$, an initial condition for the Lorenz differential equations will go to one of the two fixed points, and for $r < 1$ an orbit will go rapidly to the origin. For $r > r_c$, all fixed points are unstable, and time dependent motion must occur.

The nature of the time dependent motion depends in detail on the values of the parameters σ, b, and r. If we hold σ and b fixed, the variation in r corresponds to changing the temperature difference between the top and the bottom of the fluid. As r becomes much larger than one, the description of the actual fluid motion by the three Lorenz equations becomes increasingly

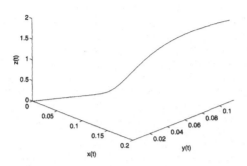

FIGURE A.1.   Time dependence of the orbit $[x(t), y(t), z(t)]$ for the Lorenz system with $\sigma = 16, b = 4$, and $r = 0.75$. The orbit goes rapidly to the fixed point at $[0, 0, 0]$.

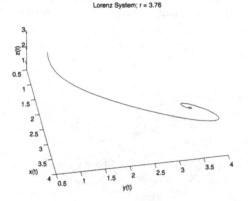

FIGURE A.2.   Time dependence of the orbit $[x(t), y(t), z(t)]$ for the Lorenz system with $\sigma = 16, b = 4$, and $r = 3.76$. The orbit goes rapidly to one of the stable fixed points.

inaccurate. Nonetheless the equations are enormously useful as a testbed for the numerical investigation of the time dependent and often nonperiodic behavior they produce.

Just to have some sense of the behavior of the orbits from these equations, let us take parameter values $\sigma = 16$ and $b = 4$. With these choices, $r_c = 19.37$. In Figure A.1 we show the result of taking $r = 0.75$ and beginning with $[x(0), y(0), z(0)] = [2.37, 4.91, 5.74]$. Clearly the orbit goes rapidly to $[0, 0, 0]$. Next in Figure A.2 we take $r = 3.76$, and the same initial condition now goes to $\pm\sqrt{b(r-1)}$. Changing the initial condition slightly and keeping r fixed also drives the orbit to one of the stable fixed points as we

Lorenz System; r = 3.76

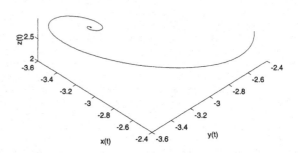

FIGURE A.3.   Time dependence of the orbit $[x(t), y(t), z(t)]$ for the Lorenz system with $\sigma = 16, b = 4$, and $r = 3.76$ with a different initial condition than used in Figure A.2. The orbit goes rapidly to the other stable fixed point.

Lorenz System; r = 45.92

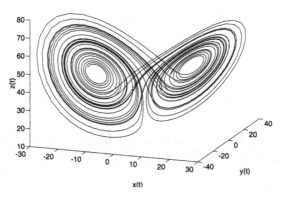

FIGURE A.4.   Time dependence of the orbit $[x(t), y(t), z(t)]$ for the Lorenz system with $\sigma = 16, b = 4$, and $r = 45.92$. The orbit goes rapidly to the Lorenz attractor.

see in Figure A.3. Finally, if we take $r = 45.92$ which is well within the regime where all fixed points are unstable, we see in Figure A.4 the familiar picture of the Lorenz attractor.

This is an example of the kind of sequence we expect to see in most physical systems as the parameter representing stress on the system is increased. More precisely what we are changing is the ratio of the stress on the system, here the temperature difference driving the buoyancy force on the fluid, to the dissipation in the system. The former drives the system to new orbital domains, and the latter damps out such motions. The balance between this driving and dissipation is the origin of the new states into which the dynamics goes when moving away from simple time independent fixed points.

Glossary

Active Degrees of Freedom The number of active dynamical variables seen in the data. This can depend on the sensor used in real experiments. Local false nearest neighbors allows one to determine this quantity in a practical manner.

Attractor The set of points in phase space visited by the solution to an evolution equation long after transients have died out. An attractor can have an integer dimension—a regular attractor—or a fractional dimension—a **strange** attractor.

Average Mutual Information The amount (in bits if base two logarithms are used) one learns about one observation from another observation **on the average over all measurements**. This is a statistic introduced in communications theory by Shannon [Sha48] which describes chaotic nonlinear systems as they are information sources.

Basin of Attraction The set of points in phase space which, taken as initial conditions for an evolution rule, lead to the same attractor. The boundaries between basins of attraction can be fractional dimensional.

Bifurcation A change in the state or topology of the solution to an evolution equation when a parameter is varied. A bifurcation is associated with changes in stability of solutions to dynamical nonlinear equations.

Cantor Set An uncountablyuncountably infinite set of points in R^d which (a) contains no subintervalsubinterval of R^d, (b) is closed, and (c) each point of which is an accumulation point. The dimension of such a set is typically factional. We have called this dimension d_A since in physical applications the Cantor set is an attractor of a dynamical system.

Chaos Deterministic evolution of a nonlinear system which is between regular behavior and stochastic behavior or 'noise'. This motion of nonlinear systems is slightly predictable, nonperiodic, and specific orbits change exponentially rapidly in response to changes in initial conditions or orbit perturbations.

Embedding Dimension The dimension of phase space required to unfold the attractor of a nonlinear systems from the observation of scalar signals from the source. This dimension is integer.

False Nearest Neighbors Nearest neighbors on an attractor which have arrived near one another by projection because the attractor is being viewed in a space of dimension too low to completely unfold it.

Flows and Maps Flows are evolution rules in continuous time. Maps are evolution rules in discrete time. Maps can come from flows by finite sampling associated with real experiments or from Poincaré sections which view a continuous orbit stroboscopically.

Fractal Dimensions The dimensions of an infinite set of points associated with the way the density of points scales with small volumes surrounding the points. Fractional values for the dimensions are typical for strange attractors of nonlinear systems.

Hamiltonian Systems Laws of evolution corresponding to closed systems or systems without dissipation. The equations of motion come from a scalar 'Hamiltonian', and when expressed in canonical coordinates, the Hamiltonian flow preserves volumes in phase space.

Lyapunov Exponents The rate at which nearby orbits diverge from each other after small perturbations when the evolution of a nonlinear system is chaotic. There are M exponents for evolution of an M dimensional flow or map. For a dissipative system, the sum of these exponents is negative. For a flow one exponent must be zero.

Phase Space/State Space Nonlinear systems are described by multidimensional vectors labeled by continuous or discrete time. The space in which these vectors lie is called state space or phase space. The dimension of phase space/state space is integer.

References

[Aba93]   H. D. I. Abarbanel. Analyzing and using time-series observations from chaotic systems. In J. M. T. Thompson and S. R. Bishop, editors, *Nonlinearity and Chaos in Engineering Dynamics*, pages 379–392, John Wiley and Sons, Chichester, 1993.

[ABK90]   Henry D. I. Abarbanel, Reggie Brown, and J. B. Kadtke. Prediction in chaotic nonlinear systems: Methods for time series with broadband Fourier spectra. *Physical Review A*, 41:1782–1807, 1990.

[ABK91]   H. D. I. Abarbanel, R. Brown, and M. B. Kennel. Variation of Lyapunov exponents on a strange attractor. *Journal of Nonlinear Science*, 1:175–199, 1991.

[ABK92]   H. D. I. Abarbanel, R. Brown, and M. B. Kennel. Local Lyapunov exponents from observed data. *Journal of Nonlinear Science*, 2:343–365, 1992.

[ABST93]   Henry D. I. Abarbanel, Reggie Brown, J. J. (SID) Sidorowich, and L. Sh. Tsimring. The analysis of observed chaotic data in physical systems. *Reviews of Modern Physics*, 65:1331–1392, 1993.

[ACE*87]   D. Auerbach, P. Cvitanović, J.-P. Eckmann, G. Gunaratne, and I. Procaccia. Exploring chaotic motion through periodic orbits. *Physical Review Letters*, 58:2387–2389, 1987.

[ACP*94]   H. D. I. Abarbanel, T. Carroll, L. M. Pecora, J. J. ("SID") Sidorowich, and L. Sh. Tsimring. Predicting physical variables in time delay embedding. *Physical Review E*, 49:1840–1853, 1994.

[AGK94]   P. M. Alsing, A. Gavrielides, and V. Kovanis. History-dependent control of unstable periodic orbits. *Physical Review E*, 50:1968–1977, 1994.

[AGLR95] H. D. I. Abarbanel, Z. Gills, C. Liu, and R. Roy. Quantum fluctuations in the operation of a chaotic laser. *Physical Review A*, 51:1175, 1995.

[AHR*95] H. D. I. Abarbanel, R. Huerta, M. I. Rabinovich, P. F. Rowat, N. F. Rulkov, and A. I. Selverston. Synchronized action of synaptically coupled chaotic single neurons: I. Simulations using realistic model neurons. 1995. Submitted to *Biological Cybernetics*, June, 1995.

[AK93] H. D. I. Abarbanel and M. B. Kennel. Local false neighbors and dynamical dimensions from observed chaotic data. *Physical Review E*, 47:3057–3068, 1993.

[AKG*94] H. D. I. Abarbanel, R. A. Katz, T. Galib, J. Cembrola, and T. W. Frison. Nonlinear analysis of high-Reynolds-number flows over a buoyant axisymmetric body. *Physical Review E*, 49:4003–4018, 1994.

[Arn78] V. I. Arnol'd. *Mathematical Methods of Classical Mechanics*. Springer, Berlin, 1978.

[Arn82] V. I. Arnol'd. *Geometrical Methods in the Theory of Ordinary Di erential Equations* . Springer , New York, 1982.

[AS93] H. D.I. Abarbanel and M. M. Sushchik. True local Lyapunov exponents and models of chaotic systems based on observations. *Int. J. Bif. Chaos*, 3:543–550, 1993.

[AVPS91] V. S. Anishchenko, T. E. Vadivasova, D. E. Postnov, and M. A. Safonova. Forced and mutual synchronization of chaos. *Radioeng. and Electron.*, 36:338–351, 1991.

[AVR86] V. S. Afraimovich, N. N. Verichev, and M. I. Rabinovich. General synchronization. *Izv. VUZ. Radiophiz.*, 29:795–803, 1986.

[BBA91] R. Brown, P. Bryant, and H. D. I. Abarbanel. Computing the Lyapunov spectrum of a dynamical system from observed time series. *Physical Review A*, 43:2787–2806, 1991.

[BGGS80] G. Bennetin, L. Galgani, A. Giorgilli, and J.-M. Strelcyn. Lyapunov characteristic exponents for smooth dynamical systems and for Hamiltonian systems: a method for computing all of them. Part 2. Numerical applications. *Meccanica*, 15:21–, 1980.

[BP85] R. Badii and A. Politi. Statistical description of chaotic attractors: The dimension function. *Journal of Statistical Physics*, 40:725–750, 1985.

[BR91] C. Bracikowski and R. Roy. Chaos in a multimode solid-state laser system. *Chaos*, 1:49–64, 1991.

[Bri90] K. Briggs. An improved method for estimating Liapunov exponents of chaotic time series. *Phys. Lett. A*, 151:27–32, 1990.

[Bro85] W. L. Brogan. *Modern Control Theory*. Prentice-Hall, Englewood Cliffs, NJ, second edition, 1985.

[Bro93] R. Brown. Orthogonal polynomials as prediction functions in arbitrary phase space dimensions. *Physical Review E*, 47:3962–3969, 1993.

[Can81] B. J. Cantwell. Organized motion in turbulent flow. *Ann. Rev. Fluid Mech.*, 13:457–515, 1981.

[Can83] G. Cantor. Ueber unendliche, lineare punktmannfaltigkeiten. *Math. Ann.*, 21:545–591, 1883.

[Cas89]   M. Casdagli. Nonlinear prediction of chaotic time series. *Physica D*, 20:335–356, 1989.

[CH92]   R. Cawley and G. Hsu. Local-geometric-projection method for noise reduction in chaotic maps and flows. *Physical Review A*, 46:3057–3082, 1992.

[CH93]   M. C. Cross and P. C. Hohenberg. Pattern formation outside of equilibrium. *Reviews of Modern Physics*, 65:851–1112, 1993.

[Cha61]   S. Chandrasekhar. *Hydrodynamic and Hydromagnetic Stability*. Cambridge University, Cambridge, 1961.

[CMN*87]   K. Coffman, W. D. Mc Cormick, Z. Noszticzius, R. H. Simoyi, and H. L. Swinney. Universality, multiplicity, and the effect of iron impurities in the Belousov-Zhabotinskii reaction. *J. Chem. Phys.*, 86:119–129, 1987.

[CP91]   T. L. Carroll and L. M. Pecora. Synchronizing chaotic circuits. *IEEE Trans. Circuits and Systems*, 38:453–456, 1991.

[Cra91]   J. D. Crawford. Introduction to bifurcation theory. *Reviews of Modern Physics*, 63:991–1037, 1991.

[CSY91]   M Casdagli, T. Sauer, and J. A. Yorke. Embedology. *J. Stat. Phys*, 65:579–616, 1991.

[DGOS93]   M. Ding, C. Grebogi, E. Ott, and T. Sauer. Plateau onset for correlation dimension–when does it occur? *Physical Review Letters*, 70:3872–3875, 1993.

[DN79]   R. Devaney and Z. Nitecki. Shift automorphisms and the Hénon mapping. *Comm. Math. Phys.*, 67:137–146, 1979.

[Dra92]   P. G. Drazin. *Nonlinear Systems*. Cambridge University, Cambridge, 1992.

[DRS90]   W. Ditto, S. N. Rauseo, and M. L. Spano. Experimental control of chaos. *Physical Review Letters*, 65:3211–3214, 1990.

[EKRC86]   J.-P. Eckmann, S. O. Kamphorst, D. Ruelle, and S. Ciliberto. Liapunov exponents from time series. *Physical Review A*, 34:4971–4979, 1986. A similar recursive method was shown to me independently by E. N. Lorenz in 1990.

[EN91]   C. Essex and M. A. H. Nerenberg. Comments on 'deterministic chaos: The science and the fiction' by D. Ruelle. *Proc. Roy. Soc. A*, 435:287–292, 1991.

[ER85]   J.-P. Eckmann and David Ruelle. Ergodic theory of chaos and strange attractors. *Reviews of Modern Physics*, 57:617, 1985.

[FBF77]   J. H. Friedman, J. L. Bentley, and R. A. Finkel. K-d tree. *ACM Trans. Math. Software*, 3:209–226, 1977.

[FC91]   T. M. Farabee and M. J. Casarella. Spectral features of wall pressure fluctuations beneath turbulent boundary layers. *Phys. Fluids A*, 3:2410–2420, 1991.

[FHB*91]   L. Flepp, R. Holzner, E. Brun, M. Finardi, and R. Badii. Model identification by periodic-orbit analysis for NMR-laser chaos. *Physical Review Letters*, 67:2244–2247, 1991.

[fra89]   Andrew M. fraser. *Information Theory and Strange Attractors*. PhD thesis, University of Texas, Austin, May 1989.

[FS86]   A. M. Fraser and H. L. Swinney. Independent coordinates for strange attractors from mutual information. *Phys. Rev. A*, 33:1134–1140, 1986.

[FS89]   J. D. Farmer and J. J. ("SID") Sidorowich. Exploiting chaos to predict the future and reduce noise. In Y.-C. Lee, editor, *Evolution, Learning and Cognition*, pages 277–330, World Scientific, Singapore, 1989.

[FY83]   H. Fujisaka and T. Yamada. Stability theory of synchronized motion in coupled-oscillator systems. *Prog. Theor. Phys.*, 69:32–47, 1983.

[Gal68]   R. G. Gallager. *Information Theory and Reliable Communication*. John Wiley and Sons, New York, 1968.

[GBP88]   P. Grassberger, R. Badii, and A. Politi. Scaling laws for invariant measures on hyperbolic and nonhyperbolic attractors. *Journal of Statistical Physics*, 51:135–178, 1988.

[GH83]   J. Guckenheimer and P. Holmes. *Nonlinear Oscillations, Dynamical Systems, and Bifurcations of Vector Fields*. Springer, New York, 1983.

[GH93]   M. Gell-Mann and J. B. Hartle. Classical equations for quantum systems. *Physical Review D*, 47:3345–3382, 1993.

[GIR*92]   Z. Gills, C. Iwata, R. Roy, I. B. Schwartz, and I. Triandaf. Tracking unstable steady states: extending the stability regime of a multimode laser system. *Physical Review Letters*, 69:3169–3172, 1992.

[GLC91]   M. Giona, F. Lentini, and V. Cimagalli. Functional reconstruction and local prediction of chaotic time series. *Physical Review A*, 44:3496–3502, 1991.

[GP83]   P. Grassberger and I. Procaccia. Characterization of strange attractors. *Phys. Rev. Letters*, 50:346–349, 1983.

[GRS84]   A. V. Gaponov-Grekhov, M. I. Rabinovich, and I. M. Starobinets. Dynamical model of spatial development of turbulence. *Sov. Phys. JETP Lett.*, 39:668–691, 1984.

[GS88]   Martin Golubitsky and David G. Shaeffer. *Singularities and Groups in Bifurcation Theory*. Applied Mathematical Sciences, Volume 51, Springer, New York, 1985-1988.

[GV89]   G. H. Golub and C. F. VanLoan. *Matrix Computations*, 2nd Edition. The Johns Hopkins, Baltimore and London, 1989.

[Ham90]   S. Hammel. A noise reduction method for chaotic systems. *Phys. Lett. A*, 148:421–428, 1990.

[HJM85]   S. Hammel, C. K. R. T. Jones, and J. Maloney. Global dynamical behavior of the optical field in a ring cavity. *Journal of the Optical Society of America B*, 2:552–564, 1985.

[Hun91]   E. R. Hunt. Stabilizing high-period orbits in a chaotic system: the diode resonator. *Physical Review Letters*, 67:1953–1955, 1991.

[Ike79]   K. Ikeda. Multiple-valued stationary state and its instability of the transmitted light by a ring cavity. *Opt. Commun.*, 30:257–261, 1979.

[KA95]   M. B. Kennel and H. D. I. Abarbanel. False strands and false neighbors. *Physical Review E*, 1995. to be published.

[KBA92]   M. B. Kennel, R. Brown, and H. D. I. Abarbanel. Determining minimum embedding dimension using a geometrical construction. *Physical Review A*, 45:3403–3411, 1992.

[KG92]   D. T. Kaplan and L. Glass. Direct test for determinism in a time series. *Physical Review Letters*, 68:427–429, 1992.

[Kim92]   H. J. Kimble. Quantum fluctuations in quantum optics–Squeezing and related phenomena. In J. Dalibard, J.-M. Raimond, and J. Zinn-Justin, editors, *Fundamental Systems in Quantum Optics*, pages 545–674, North Holland, Amsterdam, 1992.

[Kol58]   A. N. Kolmogorov. A new metric invariant for transitive dynamical systems and automorphisms in lebesgue spaces. *Dokl. Akad. Nauk., USSR*, 119:861–864, 1958. English Summary in Mathematical Reviews, Volume 21, p. 386 (1960).

[KY79]   J. L. Kaplan and J. A. Yorke. Chaotic behavior in multidimensional difference equations. In H.-O. Peitgen and H.-O. Walther, editors, *Functional Di erential Equations and Approximation of Fixed Points*, pages 204–227, Springer , Berlin, 1979.

[KY90]   E. Kostelich and J. A. Yorke. Noise reduction-finding the simplest dynamical system consistent with the data. *Physica D*, 41:183–196, 1990.

[LK89]   D. P. Lathrop and E. J. Kostelich. Characterization of an experimental strange attractor by periodic orbits. *Physical Review A*, 40:4928, 1989.

[LL91]   A. J. Lichtenberg and M. A. Lieberman. *Regular and Chaotic Dynamics* (formerly 'Regular and Stochastic Motion'). Volume 38 of *Applied Mathematical Sciences*, Springer, New York, 2nd edition, 1991.

[LM94]   L. A. Lugiato and J. Malone. Pattern formation in lasers. *Rivista del Nuovo Cimento*, 17:1–85, 1994.

[Lor63]   E. N. Lorenz. Deterministic nonperiodic flow. *J. Atmos. Sci.*, 20:130–141, 1963.

[LR89]   P. S. Landa and M. G. Rozenblum. A comparison of methods for constructing a phase space and determining the dimension of an attractor from experimental data. *Sov. Phys.—Tech. Phys.*, 34:1229–1232, 1989.

[LR91]   P. S. Landa and M. Rozenblum. Time series analysis for system identification and diagnostics. *Physica D*, 48:232–254, 1991.

[MA91]   P.-F. Marteau and H. D. I. Abarbanel. Noise reduction in chaotic time series using scaled probabilistic methods. *Journal of Nonlinear Science*, 1:313–343, 1991.

[Man81]   R. Mañé. D. Rand and L. S. Young, editors, *Dynamical Systems and Turbulence, Warwick 1980*, page 230, Springer, Berlin, 1981.

[Moo92]   F. C. Moon. *Chaotic and Fractal Dynamics: An Introduction for Applied Scientists and Engineers*. John Wiley and Sons, New York, 1992.

[MSN*91]   G. B. Mindlin, H. G. Solari, M. A. Natiello, R. Gilmore, and X.-J. Hou. Topological analysis of chaotic time series data from the Belousov-Zhabotinskii reaction. *Journal of Nonlinear Science*, 1:147–173, 1991.

[Mye]   C. Myers. Private Communication.

[ND92]   G. Nitsche and U. Dressler. Controlling chaotic dynamical systems using time delay coordinates. *Physica D*, 58:153–164, 1992.

[NS83]   R. W. Newcomb and S. Sathyan. An RC op amp chaos generator. *IEEE Trans. Circuits and Systems*, 30:54–56, 1983.

[OGY90]   E. Ott, C. Grebogi, and J. A. Yorke. Controlling chaos. *Physical Review Letters*, 64:1196–1199, 1990.

[Ose68]   V. I. Oseledec. A multiplicative ergodic theorem. Lyapunov characteristic numbers for dynamical systems. *Trudy Mosk. Mat. Obsc.*, 19:197–, 1968.

[Ott93]   E. Ott. *Chaos in Dynamical Systems*. Cambridge University Press, New York, 1993.

[Owe81]   D. H. Owens. *Multivariable and Optimal Systems*. Academic Press, London, 1981.

[Par92]   U. Parlitz. Identification of true and spurious Lyapunov exponents from time series. *Int. J. Bif. Chaos*, 2:155–165, 1992.

[PC90]   L. M. Pecora and T. L. Carroll. Synchronization in chaotic systems. *Physical Review Letters*, 64:821–824, 1990.

[PCFS80] N. Packard, J. Crutchfield, J. D. Farmer, and R. Shaw. Geometry from a time series. *Physical Review Letters*, 45:712–716, 1980.

[Pes77]   Ya. B. Pesin. Lyapunov characteristic exponents and smooth ergodic theory. *Usp. Mat. Nauk.*, 32:55–, 1977. English Translation in Russian Math. Survey, Volume 72, p. 55-, (1977).

[Pik86]   A. S. Pikovsky. Discrete-time dynamic noise filtering. *Radio Eng. Electron. Phys.*, 9:81, 1986.

[Pow81]   M. J. D. Powell. *Approximation Theory and Methods*. Cambridge University, Cambridge, 1981.

[PV87]   G. Paladin and A. Vulpiani. Anomalous scaling laws in multifractal objects. *Phys. Repts*, 156:147–225, 1987.

[Pyr92]   K. Pyrgas. Continuous control of chaos by self-controlling feedback. *Physics Letters A*, 170:421–428, 1992.

[Rab78]   M. I. Rabinovich. Stochastic self-oscillations and turbulence. *Usp. Fiz. Nauk.*, 125:123–168, 1978. English Translation in Sov. Phys. Usp., Volume 21, pp 443-469, (1979).

[Ren70]   A. Renyi. *Probability Theory*. North-Holland, Amsterdam, 1970.

[RFBB93] C Reyl, L. Flepp, R. Badii, and E. Brun. Control of nmr-laser chaos in high-dimensional embedding space. *Physical Review E*, 47:267–272, 1993.

[RGOD92] F. J. Romeiras, C. Grebogi, E. Ott, and W. P. Dayawansa. Controlling chaotic dynamical systems. *Physica D*, 58:165–192, 1992.

[Ric94]   Michael D. Richard. *Estimation and Detection with Chaotic Systems*. PhD thesis, Massachusetts Institute of Technology, February 1994.

[Ris89]   J. Rissanen. *Stochastic Complexity in Statistical Inquiry*. World Scientific, Singapore, 1989.

[RJM*92] R. Roy, T. W. Murphy Jr., T. D. Maier, Z. Gills, and E. R. Hunt. Dynamical control of a chaotic laser: experimental stabilization of a globally coupled system. *Physical Review Letters*, 68:1259–1262, 1992.

[Rob91] S. K. Robinson. Coherent motions in the turbulent boundary layer. *Ann. Rev. Fluid Mech.*, 23:241–275, 1991.

[Roe76] O. E. Roessler. An equation for continuous chaos. *Phys. Lett. A*, 57:397, 1976.

[RSS83] J.-C. Roux, R. H. Simoyi, and H. L. Swinney. Observation of a strange attractor. *Physica D*, 8:257–266, 1983.

[RSTA95] N. F. Rul'kov, M. M. Sushchik, L. Sh. Tsimring, and H. D. I. Abarbanel. Generalized synchronization of chaos in directionally coupled chaotic systems. *Physical Review E*, 51:980-994, 1995.

[RTA94] N. F. Rul'kov, L. Sh. Tsimring, and H. D. I. Abarbanel. Tracking unstable orbits in chaos using dissipative feedback. *Physical Review E*, 50:314–324, 1994.

[Rue90] D. Ruelle. Comments on deterministic chaos: The science and the fiction. *Proc. Roy. Soc. London A*, 427:241–248, 1990.

[RVR*92] N. F. Rul'kov, A. R. Volkovskii, A. Rodriguez-Lozano, E. Del Rio, and M. G. Velarde. Mutual synchronization of chaotic self-oscillators with dissipative coupling. *Int. J. Bif. and Chaos*, 2:669–676, 1992.

[RVR*94] N. F. Rul'kov, A. R. Volkovskii, A. Rodriguez-Lozano, E. Del Rio, and M. G. Velarde. Synchronous chaotic behavior of a response oscillator with chaotic driving. *Chaos, Solitons, and Fractals*, 4, 1994.

[Sal62] B. Salzmann. Finite amplitude free convection as an initial value problem–I. *J. Atmos. Sci.*, 29:329–341, 1962.

[San93] Taiye B. Sangoyomi. *Climatic Variability and Dynamics of Great Salt Lake Hydrology.* PhD thesis, Utah State University, 1993.

[Sau92] T. Sauer. A noise reduction method for signals from nonlinear systems. *Physica D*, 58:193–201, 1992.

[SB80] J. Stoer and R. Burlisch. *Introduction to Numerical Analysis.* Springer, New York, 1980.

[Sch83] G. Schewe. On the structure and resolution of wall-pressure fluctuations associated with turbulent boundary-layer flow. *J. Fluid Mech.*, 134:311–328, 1983.

[SDG*92] T. Shinbrot, W. Ditto, C. Grebogi, E. Ott, M. Spano, and J. A. Yorke. Using the sensitive dependence of chaos (the "butterfly effect") to direct trajectories in an experimental chaotic system. *Physical Review Letters*, 68:2863–2866, 1992.

[Sha48] C. E. Shannon. A mathematical theory of communication. *Bell System Tech. J.*, 27:379–423 [Part I] and 623–656 [Part II], 1948.

[Sil86] B. W. Silverman. *Density Estimation of Statistics and Data Analysis.* Chapman and Hall, London, 1986.

[Sin59] Y. Sinai. On the concept of entropy for a dynamic system. *Dokl. Akad. Nauk., USSR*, 124:768–, 1959. English Summary in Mathematical Reviews, Volume 21, p. 286 (1960).

[Sma67]   S. Smale. Differentiable dynamical systems. *Bull. Amer. Math. Soc.*, 73:747–817, 1967.

[Smi88]   L. A. Smith. Intrinsic limits on dimension calculations. *Phys. Lett. A*, 133:283–288, 1988.

[SOGY90] T. Shinbrot, E. Ott, C. Grebogi, and J. A. Yorke. Using chaos to direct trajectories to targets. *Physical Review Letters*, 65:3215–3218, 1990.

[Spr91]   R. F. Sproull. Fast *k-d* tree. *Algorithmica*, 6:579–589, 1991.

[SS85]   M. Sano and Y. Sawada. Measurement of Lyapunov spectrum from a chaotic time series. *Phys. Rev. Lett.*, 55:1082–1085, 1985.

[Tak81]   F. Takens. Detecting strange attractors in turbulence. In D. Rand and L. S. Young, editors, *Dynamical Systems and Turbulence, Warwick 1980*, page 366, Springer , Berlin, 1981.

[The90]   J. Theiler. Estimating fractal dimension. *J. Optical Soc. Am. A*, 7:1055–1073, 1990.

[Tre68]   H. L. Van Trees. *Detection, Estimation, and Modulation Theory*. Wiley, New York, 1968.

[TS86]   J. M. T. Thompson and H. B. Stewart. *Nonlinear Dynamics and Chaos*. John Wiley and Sons, Chichester, 1986.

[VR93]   A. R. Volkovskii and N. F. Rul'kov. Synchronous chaotic response of a nonlinear oscillator system as a principle for the detection of the information component of chaos. *Tech. Phys. Lett.*, 19:97–99, 1993.

[VS89]   J. Vastano and H. L. Swinney. Information transport in spatio-temporal chaos. *Physical Review Letters*, 72:241–275, 1989.

[WR90]   H. G. Winful and L. Rahman. Synchronized chaos and spatiotemporal chaos in arrays of coupled lasers. *Physical Review Letters*, 65:1575–1578, 1990.

[Yam67]   T. Yamamoto. On the extreme values of the roots of matrices. *Journal of the Mathematical Society of Japan*, 19:175–178, 1967.

Index